THE LOST TREES OF WILLOW AVENUE

ALSO BY MIKE TIDWELL

*Bayou Farewell: The Rich Life and
Tragic Death of Louisiana's Cajun Coast*

The Ponds of Kalambayi: A Peace Corps Memoir

*In the Mountains of Heaven:
True Tales of Adventure on Six Continents*

THE LOST TREES OF WILLOW AVENUE

A Story of
CLIMATE AND HOPE
on One American Street

MIKE TIDWELL

ST. MARTIN'S PRESS
NEW YORK

First published in the United States by St. Martin's Press, an imprint of
St. Martin's Publishing Group

THE LOST TREES OF WILLOW AVENUE. Copyright © 2025 by Michael Tidwell. All rights reserved. Printed in the United States of America. For information, address St. Martin's Publishing Group, 120 Broadway, New York, NY 10271.

www.stmartins.com

Designed by Kelly S. Too

The Library of Congress Cataloging-in-Publication Data
is available upon request.

ISBN 978-1-250-36226-1 (hardcover)
ISBN 978-1-250-36227-8 (ebook)

Our books may be purchased in bulk for promotional, educational, or business use. Please contact your local bookseller or the Macmillan Corporate and Premium Sales Department at 1-800-221-7945, extension 5442, or by email at MacmillanSpecialMarkets@macmillan.com.

First Edition: 2025

10 9 8 7 6 5 4 3 2 1

For Beth
No matter the weather, my sunlight

CONTENTS

Introduction   1

PART I
IN THE BEGINNING, THE TREES FELL
1. Why Are the Trees Dying?   9
2. The Worst of It   25

PART II
ONE YEAR ON WILLOW AVENUE: 2023
3. The Undertaker   43
4. Cherry Trees   69
5. An Act of God   95
6. The End of Fire   122
7. Hard to Breathe   146
8. In Sickness and in Storm   169
9. Solutions   181
10. Grave Site   209
11. Burial   241

Epilogue: December Endings   253
Postscript   263
Acknowledgments   269
Index   273

Introduction

PHOTO BY KEVIN MILLS

The biggest trees on my street weren't yet sick and dying when I moved to the 7100 block of Willow Avenue in Takoma Park, Maryland. There were true giants back then, red and white oaks, tall and broad, offering a daydream greenery of good health. I was in good health, too. I was twenty-nine years old. I ran three miles per day. I grew native wildflowers in my backyard garden a few hundred feet from the border with Washington, DC.

But that was 1991, before the chaos of climate change really settled in over this narrow block of fourteen houses—and settled in over the world.

Thirty years ago, scientists and journalists had to travel to the Arctic or Australia's Great Barrier Reef to see the beginning impacts of global warming up close. Three decades later—after the dumping of nine hundred more gigatons of carbon dioxide into the atmosphere from oil, coal, and gas combustion—those impacts are everywhere. My neighborhood block, a microdot on any Google map, has shifted to a whole new universe. Maybe you see it where you live, too.

Those old oaks, casting muffled pools of overlapping shade, formed a durable ceiling of branches here in the 1990s and into the 2000s. Then came the years of heat, the weird rain, the beetles, and finally, starting in 2019, a sudden calamity. Today, tens of thousands of mature trees across Takoma Park and adjoining cities and counties survive only as mute tombstones, the chainsawed stumps of a region-wide graveyard of lost giants.

There were no deer thirty years ago on my block; now they roam everywhere, spreading Lyme disease from ticks that survive our milder winters. Neighbors show photos of those "old" winters, decades back, with bundled-up children atop sleds in the deep snow of lower Willow Avenue's steep hill. And past summers? Old newspapers show fewer overheated people and faces, it seems, at the Fourth of July parade. Now umbrellas are a common parade feature as the natural parasols of trees decline. My church, meanwhile, a block from my house, never experienced urgent water problems until the altered weather patterns of the last few years. Now there's an elevated flood berm on one side of the church—price tag $45,000—to keep water out of the basement preschool.

Every day, I think about this: If life is so different here, so scrambled in this old trolley suburb bordering America's capital city, where can anyone hide on this planet? At one point in my long personal battle with Lyme disease, picked up from a tick in my backyard, I couldn't read or write or understand someone speaking directly to me. But I have access to medicine and professional treatment. I can't imagine the lot of Senegalese villagers penniless and suffering through the latest round of climate-induced dengue fever. Or the paycheck-to-paycheck Louisiana shrimper, buffeted by bigger hurricanes and ruined by drowning wetlands.

There are still big trees on my block today—elder oaks, shrinking in number, nobly hanging on despite it all. But if they could, if those trees had legs, they would run away from this place, I think. They would head north, maybe three hundred miles, to western New York or southern Canada, to a latitude where at least the temperature is closer to what they were born into as acorns generations ago in my Maryland town. But that's just temperature. They'd still be prone to the more frequent storms and higher winds and altered insect patterns that are stressing out trees and killing them the world over. Even with legs, the old oaks on my block could not run away.

I did not become a climate activist years ago to save trees. I did it, first and foremost, to save the future of my son, Sasha, now twenty-seven. But the oaks on my street are a fair measurement of our collective progress on global warming. Wherever mature oak trees are found, in urban forests or wilderness settings, they are a keystone species, indicating ecological health. Beyond the bounty of acorns, they are evolved to support and nourish more insects—and thus more birds and reptiles and mammals that feed on them—than any other tree genus in North America. The loss of even one big oak can radically disrupt the ecosystem of an entire local street. On my block, we've lost four giant oaks just since 2019—and others will pass away soon.

The main solution to climate change, of course, is clean energy. And on my block, just as we've seen the passing of trees, we've seen the coming of clean energy. As a nation, we've made solid strides on renewable power over the past thirty years. Low-cost solar panels hang on my roof and on commercial buildings and apartments all around me. And electric cars drive constantly up and down my street in this politically progressive neighborhood, making that pleasant humming sound my neighbor Paula calls "the faint harmony of angels."

A massive climate change bill passed by the US Congress in 2022 will effectively lock in the clean-energy revolution in this country. A complete reversal is just no longer possible no matter who occupies the White House—or what majorities make up Congress. Detroit is steadily retooling to make electric cars. Utility-scale solar power is now the cheapest energy in the history of . . . energy.

I've spent the last quarter century of my life fighting for these energy breakthroughs. I've lobbied on Capitol Hill, organized rallies in state capitals, and peacefully gone to jail blocking oil and gas pipelines. But clean energy alone can no longer save us. The genie from that bottle is out, yes, but we waited way, way too long to pop the cork. Full deployment is still many years away, which means carbon emissions will not fall at the rapid pace we need here and worldwide.

Given the lag, United Nations secretary-general António Guterres warns: "We are on a highway to climate hell with our foot still on the accelerator." That might have sounded like hyperbole just a few years ago. Now I just walk outside my front door and know he's probably right.

Already, serious policymakers are calling for "negative emissions," an attempt to suck carbon out of the air and store it in the ground in the form of crushed rock or with giant carbon-filtering machines or through agricultural practices. It's unclear whether these methods will make a guaranteed difference. They're expensive and almost certainly time-consuming beyond the planet's own stern stopwatch.

Others say we need to consider one more tool. It's a powerful tool capable of cooling the planet quickly, buying us some time. The technique is called *solar radiation modification*, or simply *solar geoengineering*. It sounds so crazy. But some really smart people at Harvard and elsewhere believe we have no choice but to explore various techniques to reflect sunlight away from the planet and back into outer space—and do it for decades. Think of it as a giant tree canopy for the whole world, a kind of solar shade high in the sky that would effectively replace—for a while—the dying oaks in my neighborhood and the wildfire-ravaged pines in California.

I've harbored doubts about this technique for a long time. But in January 2023, exhausted from twenty years of full-on climate activism while the oceans were still rising and heat waves were still blistering the planet and trees were still dying on my block, I began a quest to learn all I could about this emerging realm of "lifeboat" climate science.

This book is the story of that learning, coupled with the saga of my changing street, where human and ecological communities continue to suffer and adapt in moving and unpredictable ways.

Only in the past few years has a book like this been possible to write. Climate change has accelerated so much, with such extreme impacts everywhere, that you can now throw a dart at a spinning, lacquered globe, and wherever that dart lands, right at that tiny place, you can write a whole book about the conspicuous disruptions happening there. The story that follows—hyperlocal, replete with true narratives of life and death in one neighborhood—is proof of that. It's a harrowing and hopeful proxy for every street and every place in this nation and beyond.

Part I

IN THE BEGINNING, THE TREES FELL

1

Why Are the Trees Dying?

She probably weighed twenty tons when they finally cut her down on Willow Avenue. She had outlived Teddy Roosevelt, World War I, the Great Depression, the 1960s, the Reagan administration, September 11, the dawn of smartphones.

The old oak produced a million acorns in her life, at least. Her own existence began with a single acorn in the 1870s, likely buried by a blue jay or a squirrel and forgotten. And when she died, one arborist estimated five hundred different life-forms had inhabited her massive crown, from ravens to rat snakes to cicadas to tiny aphids to a patchwork of moss and lichen.

For more than a century, storms and droughts came and

went on the 7100 block of Willow Avenue in Takoma Park, Maryland. New houses were built, others burned down. And this southern red oak (*Quercus falcata*) was always there, the emerging giant just behind the backyard porch of what eventually became the Miller property. Children climbed on her branches for generations. Birds made nests. There were lightning strikes.

Before so much changed on this block—before the shocking rain of 2018 and all that followed—I remember walking home on hot summer days from the DC subway. I'd walk past broiling urban parking lots and shade-free DC storefronts until I reached Willow Avenue, the Maryland border, and there was the Miller Tree.

By herself, she shaded a vast portion of the block, a sentry at the edge of a magnificent urban-suburban forest covering much of Takoma Park. A drop in temperature of nearly ten degrees came with her shadow. That shadow poured down from the sixty-foot-tall trunk and a canopy of fully leafed branches.

Those branches stretched languorously to the southwest, nearly touching a giant walnut tree bordering the back lawn of the Pande-Gordy house. To the northeast, the Miller Tree stretched to within conversation distance of a big white oak at the Kurtz-Greenberger house. Those neighboring trees, in turn, reached toward others—and so on throughout much of this municipality of eighteen thousand people. Over its history, so obsessed has the city of Takoma Park been with planting and caring for trees that a big share of its streets are named for them: Maple, Holly, Tulip, Cedar, Dogwood, Birch, Willow—all avenues.

But the big oaks dominate here and are highly valued—and no wonder. People describe a sense of well-being, a shift in mood when standing under big trees, including oaks. It's not an illusion. Studies show a person's blood pressure typically drops while "forest bathing." Just seeing a tree through a window can help patients recover faster from sickness. Your immune system improves. Whatever worries or ails you, you feel better near a big tree.

Lisa and Dave Miller bought the house at 7120 Willow Avenue in the autumn of 2010. Eager to start a family, they obsessed over the handsome oak out back, feeling her giant spirit hover protectively over the entire backyard. Dave, the family grill chef, envisioned an outdoor smoker for brisket and beer-can chicken. Lisa came from a family of hunters and fishermen. Her grandfather had owned a charter fishing boat on the Chesapeake Bay. She craved the outdoors and all its natural treasures.

Those treasures, of course, were originally home to a far different people in this region: the Anacostan people of the Piscataway tribe. They lived on migratory shad from the nearby Potomac River and—in lean years—a flour made from dried oak acorns. By the mid-1600s, that world was already slipping away, lost to the axes and guns of invading European settlers.

For the next two centuries, the area that is today Takoma Park was timbered and tilled by scattered farmers. Then in the 1880s and '90s, Willow Avenue and surrounding streets were mapped out as a trolley suburb of Washington, DC. In 1905, the Miller house was constructed with newly modern efficiency, the precut and sorted materials arriving by mule wagon from the nearby Baltimore and Ohio Railroad.

In 2010 when the Millers moved in, the giant red oak in the backyard was—astonishingly—in near-perfect condition. Unlike most trees her age, she was not shedding many branches, large or small, to the ground below. She survived the massive derecho storm of 2012—a line of thunderstorms on steroids that took out many younger trees in this town. When the Millers' son, Wesley, was born in 2013, they put his crib on the second floor, in the corner room closest to the tree's massive girth, the ultimate parental vote of confidence. That confidence lasted five more years.

By chance, around the time I moved to Willow Avenue in 1991 and fell in love with its trees, scientists began to unravel some of the inner workings and mysteries of these magnificent plants. Western societies for centuries have assumed trees were more or less mute objects—giant and useful and pretty to look at but the opposite of dynamic. But in the early 1990s, using radioactive isotopes pumped into the roots, researchers discovered trees are amazingly social creatures. They talk to one another underground, learn from one another, store knowledge, and live as families whenever they can, children next to parents.

The complex root system—interwoven with an even complexer network of symbiotic fungi—makes this possible. So extensive is the underground sharing of information through chemicals and even electrical pulses that many scientists call it "the wood wide web." Trees exchange sugar and other nutrients with each other. Mothers "nurse" their young with supplemental food passed root to root. All the while, aboveground, trees are talking, too, using scent and color to warn of attacking insects or unfolding drought.

More research has shown that trees like to sleep, they can tell time, they can taste. And almost certainly, they feel pain.

In fact, many foresters and arborists believe that when the pain becomes so great inside a tree's body—as during droughts, for example—the tree starts to scream. The trunk vibrates softly in the plant's version of vocal cords. Sonic tests have confirmed these vibrations inside distressed trees.

And inside Takoma Park, Maryland, sometime in 2018 or 2019, after months of pummeling record rainfall and then weeks of drought, it seems clear that that's what was happening to thousands of old oak trees here. Their bodies had begun to shake. Their trunks were vibrating.

They were screaming.

IT WAS ONE of the strangest weather patterns observed in the twenty-first century. The jet stream—that powerful global current of air—had dipped all the way south to the Gulf of Mexico by the middle of 2018. It abandoned its normal North American summer route across lower Canada. In the process, it pulled tons of moisture up from the Gulf, hurling it toward the mid-Atlantic.

By July, weather experts were calling it an "atmospheric river." There was so much moisture, flowing so fast, it was like a river in the sky, roaring inland. Every few days, the bottom would just fall out and the river crashed down upon the Eastern Seaboard.

From mid-May to early June, DC's Ronald Reagan Washington National Airport recorded a soaking ten inches of rain. Forty miles to the north, in Ellicott City, Maryland, ten inches of rain fell *in two hours* on May 27. A ten-foot wall of water barreled through the downtown, destroying almost everything. Watching the footage, you wonder: How can so much water crash into so many buildings so quickly? One National Guardsman was killed. Meteorologists called it a "one-thousand-year flood." It was the second one-thousand-year flood in two years in that same small town.

On Willow Avenue, my twenty-one-year-old dogwood tree was the first tree to die on the block. It was planted the month my son, Sasha, was born in May 1997. It drowned in the front yard, its roots waterlogged and rotten.

Across the street, the Millers' basement flooded as never before. The ground was completely soaked when a staggering four more inches of rain fell on July 21. There was nowhere for the liquid to go. Mud and debris overwhelmed an exterior drain, sending water into their basement—three inches deep—before Dave could start the cleanup.

There are different ways to explain what happened in 2018 in the DC region and beyond. Meteorologists refer to wind flows and land-ocean temperature contrasts, making it seem almost normal. But the truth is, it was just our turn.

Against the daily background noise of climate change, epic events still astound us each year. They come more frequently now, each with a fresh jolt of warning, a new depth of sadness, and—it feels—a force bordering on evil: the sudden mass bleaching of coral reefs across half the Caribbean Sea; the collapse of Antarctic ice shelves the size of large American cities; the arrival of hurricanes like Katrina.

What happened in the mid-Atlantic starting in 2018 has gotten much less attention despite its historic scale. Nine eastern states and the District of Columbia set records for annual rainfall in 2018, most by wide margins: Delaware, Maryland, Massachusetts, New Jersey, North Carolina, Pennsylvania, Tennessee, Virginia, West Virginia. Maryland got two *feet* more rain than normal. It was so wet for so long that a big chain reaction was set in motion.

For the big oaks in my neighborhood, it began deep underground, at the foundation of tree life. It began with the roots—and with a mold called *phytophthora*.

Found naturally in the local soil, phytophthora is a water mold of the oomycetes type, always waiting for the right conditions to expand. It's not a fungus or bacterium, although it's sometimes called "fungus-like" with its own evolutionary tract. Tellingly, the name *phytophthora* translates from Greek as "plant destroyer." In the summer of 2018, nourished by the soaking deluge, it began latching onto one of its favorite hosts: the hairlike root tips of oak trees.

Phytophthora swarms over these tiny fine roots, feeds on them, and kills them. It severs almost all connection between the tree and the beneficial fungi and minerals in the soil. It limits the flow of water up into the trunk. Though you couldn't see it, by October of that year, thousands of damaged trees in Takoma Park and surrounding neighborhoods were already tapping into emergency supplies of carbohydrates stored in their trunks. As they dropped their autumn leaves that year, many were headed into the last winter of their lives.

DAVE AND LISA MILLER wound up calling three different arborists to examine their ailing tree. The giant canopy just didn't leaf out properly in the spring of 2019. A third of the tree's branches were naked and dying. Puzzled arborists initially scratched their heads and said some version of "It's an old tree." They didn't know yet that phytophthora had gone wild underground.

By June of that year, at 7128 Willow Avenue, Michele Kurtz and Scott Greenberger noticed their backyard giant was also acting strange. It was a white oak—a tree so regionally common and long-lived that it is the official state tree of Maryland. Throughout June, it shed branches in a torrent.

By the end of the month, big swaths of bark began falling off its trunk.

Suddenly, neighborhood LISTSERVs lit up with dire subject lines that summer of 2019. "Is my tree dead?" "So many oaks are turning brown!" "Has anyone found answers?" City officials eventually organized a town hall meeting titled "Disease and Decline: Why Are the Oak Trees Dying?" The crowd was standing room only.

Commercial tree companies were flooded with calls across the DC region by midsummer. Their response was pretty much the same at first: "We don't know why so many oaks are getting sick at the same time. We've never seen this before."

The phenomenon went far beyond DC. The biggest tree companies—Bartlett, SavATree, Adirondack—reported variations of this 2019 story from North Carolina to southern New England, from Delaware to West Virginia. Although mortality was occurring in forests, it appeared worse in urban and suburban landscapes.

Humans weren't the only ones noticing that something was amiss with the oaks. If the first explosion in my neighborhood was a mold underground, the second was a tiny, hard-shelled insect up above.

Ambrosia beetles in the mid-Atlantic—representing two subfamilies of weevils—have one thing in common: highly sensitive, feathery antennae that can "smell" from far away the chemical distress signals coming from damaged trees. One of their favorite scents is ethanol, a compound commonly vented by oak trees when they are in deep distress. Even at trace levels in the air, ambrosia beetles can detect ethanol. They are expert hunters, bloodhounds on six legs.

At some point that summer in my neighborhood, the first ambrosia beetle began probing and drilling into the first oak tree. By definition, that tree was already under stress, emitting volatile compounds like ethanol as a by-product of the desperate work of rebuilding destroyed roots below while maintaining new leaves above. That first tree probably put up a good fight. It flooded the beetle wound with a store of bitter tannin and other defensive phenolic compounds while sending out an airborne chemical message to friendly insects—predators who like to eat ambrosia beetles—telling them a meal was spreading across its trunk.

This chemical warfare between big oaks and beetles varied from tree to tree. But soon that summer, the beetles began to overrun the field. On most blocks, at least one giant oak—and often four or five—was soon exhibiting the yellowish marks of a guaranteed death sentence: fine sawdust pouring out of beetle-drilled holes all along its lower trunk.

The name *ambrosia* comes from the fungus the beetles bring into the tree, depositing it below the bark for their larvae to eat. It is nectar for the coming offspring. During optimal conditions, as in the summer of 2019 in Takoma Park, these beetles have the reproductive potential to create a new generation every twenty-one days, going from one hundred thousand to one million to one billion beetles in two months, obliterating more and more trees.

What began with a few homeowners noticing browning limbs in April was by September something entomologists called "a mass host mortality event."

IT TOOK TWO days to cut down the Miller Tree. She came apart in sections, limb by limb, at the hands of chain saw operators in hard hats swinging from ropes. A five-story crane was brought in for the trunk and the largest branches. The crane weighed ten tons, nearly crushing the Millers' driveway. It blocked all street traffic and required the removal of power lines.

The Millers sent their eight-year-old son, Wesley, to school those two days. He missed, blessedly, much of the worst trauma: the thud of the massive trunk as it was lowered to the street, the endless mechanical shriek of limbs being fed into the knives of a wood chipper.

When it was over, the Millers went into a deep state of grief from which they have yet to return. "I can't imagine losing a child," Dave said to me later. "But this loss, it totally felt like a family member."

Their tree didn't die in 2019 like so many others that summer. She beat back the beetles for a while and hung on even through the freakishly sudden flash drought that fall. After gushing for so long in 2018 and 2019, the rain spigot stopped completely for much of late summer and early fall 2019. Temperatures simultaneously soared into the nineties across Maryland—in October! Pumpkin harvests failed.

During the period of late 2019 and most of 2020, many tree companies, like Bartlett and Adirondack, deployed all their trucks, all their cranes, all their crews, all their chain saws, all their chippers, all the time—all across the DC region. It was like the mobilization after a hurricane, one arborist said. When those companies maxed out, wildcat crews showed up—the amateurs who arrive in the aftermath of natural disasters—taking down trees with rented saws, spools of rope, and pickup trucks.

For weeks on end, men in hard hats dangled from trees seemingly everywhere in Takoma Park, far above the ground, almost every day. It took miles and miles of rope to bring down all the dead trees. The "ground men" sent food, water, and saws up to the "climbers" using slipknots, timber hitches, and Blake's hitches. The climbers wore Black Diamond harnesses with aluminum carabiners and other state-of-the-art climbing equipment—plus carried the heavy saws. They used trigonometric equations to make sure each sliced piece of wood swung just the right way using ropes and pulleys or the direct cable of the crane.

Official records show we lost nearly 1,200 trees in Takoma Park during the period from mid-2019 to mid-2021, the biggest share being oaks. That's a lot for a small town. But in terms of canopy cover, it's much, much worse, given so many of these trees were spatial giants. Permitted tree removals in the city averaged about 170 per year in the dozen years prior to 2019, then jumped to a *600-per-year* average from mid-2019 to mid-2021 during the worst of the insect invasion. Since then, the average is still well over 400 per year, more than twice the pre-2019 average.

On the 7100 block of Willow Avenue, we lost three of our biggest trees in just two years. Michele and Scott's backyard white oak came down in October 2019. The branches were chipped into mulch. The trunk—they were told—was sent to a lumberyard. Not long after, Lezetta and Lin Moyer's white oak came down, riddled with beetles, near the corner of Willow and Tulip Avenue.

And finally in the spring of 2021, the Miller Tree surrendered. Barely a third of her branches leafed out that April. Her root system was clearly in tatters.

Sickening. Traumatizing. Depressing. Shocking. Appalling. These are the words used by tree experts in this region—the private arborists and academics and public works bureaucrats—when they describe the oak mortality of the past few years here. And these are the otherwise detached professionals who've seen it all. For the rest of us in places like Takoma Park, us locals, the graveyard of tree stumps is unforgiving. Chris Larkin, a senior arborist at Bartlett Tree Experts, told me, "When a tree dies in a national forest, it's one among thousands. When it dies in a backyard, it's a friend."

Every day, the Millers wake up and their majestic tree, born in the 1870s, is not there. Every day, they don't see her branches and her shade outside their kitchen window. Every day, they have proof that something has shifted on their block, that a threshold has been crossed in their city, and that the planet just doesn't seem able to support big trees anymore. The trees are dying in sustained waves, reshaping landscapes, and not just here: The giant cedars of Lebanon are burning. The fabled stone pines of Rome are succumbing to invading insects. The bristlecone pines of Utah are dying from drought.

No wonder there's so much climate anxiety in my neighborhood, such a common feeling of gloom even as electric cars and solar panels spread. There are families on almost every block in Takoma Park, like the Millers, grieving right now over close arboreal friends lost since 2019. And the rest of us live right next door or walk by the remains daily.

Part of me wonders: Could we hear them while they were dying? Could we hear, at a subconscious level, the "screaming" vibrations of trunks during stretches of drought? Could we hear the pain, too, of big oaks as they declined in extreme rain?

And are we grieving not just for the trees but for what they obviously represent—the passing of our entire planet? Are we hearing, too, consciously or not, the cries of all native life-forms around us—the birds, the insects, the smaller plants? And does this cascading cacophony affect *us*, weakening *our* immune systems, making *us* more vulnerable to diseases like Lyme? If studies show that just being under a big tree makes us happier and improves our immune systems, what happens when those trees are gone?

Another Takoma Park study, released in 2022, showed our town has suffered a net loss of forty-five acres of tree canopy in recent years, accelerating from 2018 on. And it is only a matter of time before the next atmospheric river barrels toward the mid-Atlantic. That street-level process ending with beetle-drilled holes and terrible sawdust oozing from the city's trees—it begins with warmth. A warmer atmosphere holds more moisture. A warmer atmosphere evaporates more of the world's ocean water into the sky. And at some point, all that water comes back down. It comes down—increasingly, science shows—in bursts of record precipitation.

Unless we explore new ways to dial down that heat, it won't be long before every rainfall year is a 2018 year in the mid-Atlantic. Every year will be perfect for phytophthora, the underground mold, exploding beneath our feet. The plant destroyer.

EACH SEPTEMBER, WHEN the summer thunderstorms slow down and autumn has not yet fully arrived, a magical wind comes to the 7100 block of Willow Avenue. It's a steady wind, blowing all day, sometimes for two or three days, swishing through the trees. It makes the leaves speak.

So comforting is this wind, you can almost imagine you're somewhere else, that the ocean is at the end of the block, its breeze flowing across the surf and into the trees. But it's just the changing seasons here. My old neighbor Jack calls it *September kite weather*, the early autumn companion of March, with leaves on the trees to applaud another year.

For thirty-plus years, I've looked forward to these vivid, bright, windy days in mid-September. Back in the 1990s when I first moved here, the sound was different, though. It was louder. The trees were ubiquitous, the leaves everywhere, creating a pleasant roar. You had to politely yell to neighbors on the street, yell across backyard fences, raise your voice on sidewalks.

Today, if anything, the September wind is stronger. But with fewer trees and fewer leaf collisions overhead, the sound has a softer quality now. New pockets of sound have emerged. Deeper tones. Over gaps in the tree canopy, the wind passes now like human breath over the tops of empty bottles.

Sixty years ago, the writer Rachel Carson warned that human damage to the environment was changing the very sound of our seasons. Carson, a Maryland resident for much of her adult life, observed that fewer songbirds were arriving each spring across American farms and cities, killed off by DDT, which was later banned. Her book *Silent Spring* helped launch the modern American environmental movement.

In Takoma Park, autumn is not yet silent. There are still a lot of trees here. A visitor who doesn't know our recent history will still be impressed by our canopy. On my block, on Willow Avenue, a few old oaks still inhabit the backyards to the west, and an impressive string of nine willow oaks (forty to eighty years old) line one sidewalk. But residents like me can walk you to all the stumps—or the sunken ground where stumps once were—strewn across our block and on streets citywide. My friend Daryl Braithwaite has lost five giant oaks in her one yard on Hickory Avenue in recent years, including to swarming beetles in 2019. Walk our streets and you will see the gaps. You will see why September has a different sound today.

Daryl, ironically, oversees all things trees in Takoma Park as director of the city's Public Works Department. She's greatly expanding the city's tree-planting efforts, with replacement oaks and other varieties going into the ground on public and private property. But it's unclear how big these trees will ever get or how long the mature oaks we still have will continue to live.

Chris Larkin, the arborist, told me, "Many of our native trees just aren't native anymore in this region. Our climate has changed. The trees are foreigners now, hanging on. That's why so many are dying."

A lot of locals have apparently gotten the message. One of the most common oak trees being planted here now, according to city data, is a swamp white oak, a tree native to this region but now much more popular. Some researchers at the University of Maryland are reportedly recommending it because, well, we're apparently becoming a region-wide swamp.

But I keep rooting for the other native oaks here, declining in number but still being planted in my neighborhood. One day, I hope, our climate will become native again and my son and his future kids, when they visit Takoma Park, will have to yell outside to be heard on those special days in September.

In the meantime, in the Millers' backyard, in the large gap

that now brings a downpour of sunlight, a catalpa tree has sprung up on its own. It's a fast-growing "pioneer" tree that thrives in such spaces. It stands side by side with a young American elm, barely noticeable before but now rising to take the place of its predecessor, that big old red oak, *Quercus falcata*.

2

The Worst of It

One night on Willow Avenue, when my son, Sasha, was six years old, a journalist called me on deadline. It was after dinner, and Sasha was playing with LEGOs on the living room floor, finishing an elaborate house. He was fastening LEGO "solar panels" to the roof. He thought all houses in America had solar panels since that's all his dad talked about.

The reporter was calling about a highly disturbing study—published that day in 2004—showing Antarctic ice melt could be happening much faster than previously understood. He was looking for a reaction quote.

"Step one is clear," I said as I always did in such conversations. "We must prevent the worst-case scenario from happening. We have to get off fossil fuels as fast as we can so Antarctica doesn't melt."

Yes, yes, the reporter said, but what if the worst *does* happen? What if the Thwaites Glacier fully collapses? Well, I said, it would be a full-on disaster. A third of New York City would disappear. Most of downtown London, too. Bridges would collapse in DC. There would be mass migration. And so on.

Only after I hung up, exhausted from the reporter's grilling, did I remember my dear little boy was in the room, having heard everything. "Daddy," Sasha said, LEGOs still in hand, "are we all going to die?"

This child was the whole reason I became a climate activist. This child—who would go on to play endless games of tag in the backyard and become an Eagle Scout and pitch for the local high school baseball team and teach his US congressman, Jamie Raskin, how to throw a curveball on the Willow Avenue pavement—this child was why I changed my life in mid-career, formed a nonprofit, and fought for climate solutions every day.

"No," I said that winter night. "We're not going to die." I got down on my belly, propped my chin in my hands, and looked straight at him. "I'm working with some really, really great leaders, and we're going to fix this problem. I promise."

And I meant it. Two decades ago, in the early days of the US climate movement, activists like me understood our mission. We couldn't stop global warming completely. Society had put too much carbon dioxide in the atmosphere for that. But we could "prevent the worst impacts from occurring." We always added some version of that line. By switching to clean energy fast, we were going to "avert the biggest consequences" or "mitigate the most extreme scenarios" of global warming.

And that would be a good-enough world. It would be a planet of two feet of sea-level rise, not twenty; of longer heat waves but not civil wars triggered by them; of bigger floods and longer droughts but also resilient, stable societies for Sasha's generation. It would be a world where big trees can still exist on Willow Avenue.

There was no serious talk back then of sucking carbon out of the air or reflecting sunlight away from the planet. Why would we need cooling mirrors in California's Central Valley to grow tomatoes or sulfur in the stratosphere to calm hurricanes? We would do all that with wind and solar power and electric cars—enough to stabilize the climate and phase out fossil fuels for good, the root problem. We just needed to move fast, passing clean-energy laws at the state and federal levels *now*.

In 2002, I founded the Chesapeake Climate Action Network to greatly speed up this revolution in my region. At the time, the Sierra Club and Greenpeace and the League of Conservation Voters—they all had global warming on their lists of things to do, but none had it as *the* priority.

I had read Bill McKibben's book *The End of Nature* years before, in 1990, and was gut punched by his seminal description of the dangers of the "greenhouse effect." But for a decade afterward, too busy with my career and raising a kid, I did nothing. Then came the early 2000s, when each month seemed to bring a terrifying new scientific study on climate. I could no longer look at my LEGO-crazed son and keep doing nothing. I remember being angry at the world for foisting this abrupt change on my life. None of my friends were dismantling their worlds to address the climate crisis. Students weren't rioting on campuses. But then I realized, *Mike, if* you're *not out there protesting, how can you get mad at anyone else?*

So I walked away from my treasured career as a freelance print journalist. Since my early twenties, I had traveled the world for publications like *The Boston Globe*, *National Geographic Traveler*, and *The Washington Post*. I had paddled the Amazon, ridden Silk Road horses, and backpacked across Sicily. But in June 2002, at age forty, I called all my editors and told them I was launching a nonprofit to fight climate change. They should stop calling me, I said.

When I finished making the last call, still shocked by what I was doing, I had no idea I was about to model some pretty strange behavior for the father of a grade-schooler. I was about to get arrested—many times—and put in jail.

WAY BACK IN 1965, as the Vietnam War rumbled and Gemini astronauts pioneered a path to the moon, science advisors to President Lyndon B. Johnson produced the nation's first top-level government report on global warming.

It was an amazingly accurate document, predicting a dangerous 25 percent increase in atmospheric CO_2 by the year 2000. That's almost exactly what the National Oceanic and Atmospheric Administration actually observed by that date.

"Man is unwittingly conducting a vast geophysical experiment," the report warned Johnson. The president in turn warned Congress in a follow-up letter, saying the modern world "has altered the composition of the atmosphere on a global scale . . . from the burning of fossil fuels."

But the report didn't recommend actually cutting fossil

fuels. That was unthinkable, apparently, in 1965, given the world's fast-deepening addiction to this miracle fuel. Instead, backed by complex mathematical calculations, the authors proposed creating "countervailing climatic changes" on Earth. They proposed chemically modifying high-altitude cirrus clouds and floating billions of buoyant reflective particles in the oceans. These steps would increase the planet's albedo, its reflectivity, and cool the earth.

The first official US government report on climate change was an argument for geoengineering.

Years later, I, too, visited the White House to discuss the intricacies of climate change with the president. But this time in the form of handcuffing myself to the White House fence. It was 2013, and I was there with forty-eight other activists—and stopping fossil fuels was *exactly* why we were there. A surreal pipeline had been proposed to transport tar sands oil from Alberta, Canada, all the way to Texas and to world markets beyond. Using computers much more powerful than those in the 1960s, scientists warned the Keystone XL pipeline was a "carbon time bomb" that could push the planet over the edge.

Like most world leaders by this time, President Barack Obama had declared his commitment to fighting global warming. But he had yet to exercise his power to cancel the Keystone pipeline. So with church pastors and Nebraska farmers and college students, I zip-tied myself to the White House fence and ignored police commands to leave.

In short order, we were led to paddy wagons, where I found myself seated next to the legendary civil rights leader Julian Bond. I grew up in Georgia in the 1960s, where he was one of my heroes. With time to kill, there in the paddy wagon, Bond began describing the first time he got arrested as a Black activist. It was at a lunch counter in Atlanta in 1960, at city hall. "We asked to see the manager, and when he came out, we pointed to the cafeteria sign overhead," Bond said, gesturing upward with his handcuffed wrists to re-create the moment. "The sign says, 'A cafeteria for all people.'

"'Well,' the manager responded, 'we don't really mean it.'

"'Well, we're going to sit here until you do,'" Bond recounted. Laughter filled the paddy wagon.

For the last twenty years, the US climate movement has been, in large part, about making politicians and energy companies mean it. They all said fighting climate change was important, but could they please just build this one gas pipeline over here and this expanded coal plant over there and this fracking rig right here?

Hell no, we said. And we sat down. This was a planet for all people.

I don't know how many times I've been arrested as a climate activist. Every time I make a list, I realize I've left something out. The first time was outside a coal-fired power plant in Dickerson, Maryland. We blocked the entrance so workers couldn't get in. A rabbi arrested with us told a gaggle of reporters that burning coal simply had no place in God's creation anymore.

Nonviolent civil disobedience: it's a powerful tool. I've been arrested outside the governors' mansions in Maryland and Virginia and inside the halls of Congress. Once, in 2019, I joined a group called Shut Down DC—and that's what we did. We strategically blocked a dozen key intersections with our bodies and essentially shut down—for a few hours—the DC morning commute. Our message to frustrated but largely sympathetic drivers: we're on a highway to hell unless Congress takes big action on climate change.

Fossil fuel executives like to call us radical extremists. Really? What could be more radical than knowingly altering the chemical makeup of the entire sky, combusting oil, coal, and fracked gas until disastrous atmospheric rivers come pouring down on places like Washington, DC, itself?

And extreme? What could be more extreme than something called the Atlantic Coast Pipeline? For years, the Chesapeake Climate Action Network and our allies fought this monstrosity proposed by Dominion Energy, Virginia's main power utility. The $6 billion pipeline, six hundred miles long, would transport gas from the fracking fields of West Virginia, across Virginia, to the coast of North Carolina. It would be the world's first such pipeline engineered to go almost straight up and straight down a slew of steep mountains, crossing the Blue Ridge and Allegheny ranges.

Along the way, in Bath County, Virginia, on the property of Bill and Lynn Limpert, the pipeline would destroy an entire virgin forest of four-hundred-year-old trees. Those trees—basswoods, sugar maples, hickory oaks—stretched across what locals called Miracle Ridge. It was the largest stretch of old-growth forest in the entire state. There was nothing like it even in Shenandoah National Park.

The Limperts publicly pleaded for sanity and invited opponents like CCAN to come help stop the injustice. So we came. All indications were the bulldozers would arrive in the summer of 2018. We formed a small activist city that summer, camping in shifts for weeks under and around the massive, magical trees—rain or shine. A carpenter drew up plans for treehouses that we would build quickly the minute the chain saws arrived. We would occupy those platforms as long as we could, holding out against the Dominion hard hats below.

But weeks went by, and in late September, the bulldozers had still not arrived. They never came. Activism like this all across the mountains was pounding away at Dominion's image. And court cases—filed by CCAN and a half dozen other groups—were slowing down the permits needed to continue construction.

Finally, in July 2020, Dominion made a stunning announcement: it was canceling the Atlantic Coast Pipeline. All construction suddenly stopped. Goliath had fallen. David had won. And Miracle Ridge and the miracle trees are still there in Bath County today.

And that coal plant back in Dickerson, Maryland? It was shut down and decommissioned in 2020. There is talk of turning the property into a solar farm. And on January 20, 2021, his first day in office, President Joe Biden canceled the Keystone XL pipeline for good. Tar sands oil does not—and may never—flow to the Gulf of Mexico.

By any definition, by the start of the 2020s, we were winning as a movement—at least in persuading most Americans that most fossil fuels needed to stay in the ground. I knew this in part by looking down from my seventh-floor CCAN office, a perch overlooking Takoma Park and much of DC and distant Virginia. From here, you can see heavy freight trains pass by on the old B&O line two blocks away. For years, just past 10:00 a.m., every single day, a half-mile-long coal train would rumble by, always rolling south, from where and to where we never figured out. I counted seventy-five cars some days, rolling through and past our declining urban forest, past big trees destined to die soon, the stark connection between climate cause and climate outcome visible in a single surreal frame.

But in recent years, those coal trains have gotten fewer and fewer—and I rarely see them now from my office window.

It wasn't enough, however, to be phasing out coal and stopping new pipelines. To really win, all the rooftops of those houses and businesses, also visible from my office window, stretching across America's capital, needed to be covered in solar panels and clean-energy heat pumps.

And that meant going back to the White House. To keep my long-ago promise to my son, Sasha—a promise to beat back climate calamity and reduce sea-level rise and help save the trees in his neighborhood—I needed to visit the president's house again, not to handcuff myself to the fence but as an invited guest to the Rose Garden, there to celebrate passage of a transformative bill. For the US climate movement, this had always been the ultimate goal: pass a giant federal clean-energy bill that would finally change the country and the world.

IN 2002, WHEN Sasha was still in his T-ball glory days, the Chesapeake Climate Action Network launched its very first statewide legislative campaign—a clean-electricity bill in Maryland. Back then, Sasha loved circling the T-ball bases amid the chaos of missed grounders and bad throws. His teams averaged ten "home runs" and twenty errors per game. And that's what lobbying in Annapolis, the state capital, felt like to me that first year of legislative advocacy, in 2002. I had no idea what I was doing. I was CCAN's only employee, and the fossil fuel lobbyists crushed the legislation like a bug. We lost the next year, too.

But the third year, 2004, CCAN had three employees and

a rapidly growing database of volunteers across the state. And we finally passed our bill mandating a modest 7.5 percent of the state's electricity come from clean sources like wind and solar power.

We took additional small steps for several years until 2009, when we pushed Maryland legislators to adopt a transformative bill capping CO_2 emissions from coal- and gas-fired power plants. Sasha was a preteen skateboarder by then, wearing Vans tennis shoes and playing drums in a charmingly terrible rock band whose signature song was "Homework Sucks." That same year, a federal bill to cap carbon emissions nationally died in its tracks. While the states made progress, federal energy policy was still stuck at a sixth-grade level.

Then came 2013, one of the best years of my life. Sasha became an Eagle Scout that year, capping a childhood love of camping and hiking. He had survived yellow jacket swarms in Rock Creek Park and the wilderness survival class taught by his dad for local Troop 33. His Eagle project was placing "No Dumping" signs on all 150 street drains in our city. They are still there today.

And 2013 brought a huge leap forward for the clean-energy movement on the East Coast. Legislators in Annapolis passed the Maryland Offshore Wind Act, a pioneering step in what is now a multibillion-dollar industry emerging from Massachusetts to North Carolina with the promise of tens of thousands of new jobs. By the mid-2030s, if all goes as planned, there will be enough electricity from Atlantic Ocean windmills to power every household in Maryland and much of the East Coast. Across the country by the 2010s, state policies for wind and solar power and energy efficiency—especially in blue states, but also in windy places like Iowa, Kansas, and West Texas—

were finally starting to match the scale of the problem. Energy policy in America was growing up.

Way back in the T-ball era, when Sasha was a grade-schooler, it took CCAN three years to pass a miniscule bill requiring less than one-tenth of Maryland's power come from wind and solar. Coal was still king—all across the region—and people hardly knew what a hybrid car was, much less electric. But by the time Sasha finished college in 2019, graduating from the University of Maryland, everything was in full-tilt change. He wore a black gown and mortarboard and carried across the graduation stage a degree from the school's environmental science and policy program, itself a fairly recent creation. The LEGO toy world of his childhood imagination, where every house had solar panels, was not quite here—but it was on its way.

What was still missing, nearly sixty years after Lyndon Johnson first declared climate change a world problem, was that return to the White House. We needed a federal bill to lock in and greatly expand these state successes and help transform the global energy economy. And we needed it now. The early 2020s, many scientists believed, represented our last chance to slow down the massive harm being done to our unstable planet.

WHEN IT FINALLY came, when the national climate movement finally had its crowning moment in late 2022, it was the most beautiful day ever in Washington, DC. The September sky was a poet's dream of deep blue—and a mild, dry air had chased away Washington's famous summer humidity.

I took the Metro from Takoma station, hopping off at Metro Center and walking the last four blocks to the White

House. In thirty-plus years of living here, I had never been inside the White House gate, even as a tourist. And as a professional activist, how would I have come here? It still seemed impossible, but no president had ever signed a major climate bill into law before. Ever.

I joined the line of guests along Fifteenth Street, bordering the presidential grounds. We passed through a ten-foot iron gate, then a line of metal detectors, and then I was there, on the White House South Lawn, with the Rose Garden in the distance. It was as lovely as advertised, all vivid colors: bright green lawn rolling up to bright white mansion rising up to bright blue sky. People were everywhere. Someone handed me a commemorative set of aviator sunglasses, the kind President Joe Biden likes to wear. I had waited decades for this sun-filled, dreamlike moment of people and music and celebration. So I put the glasses in my suit pocket, not wanting anything to filter this view.

It took eighteen months of ferocious coast-to-coast campaigning to pass the Inflation Reduction Act of 2022. Even after conservative Democrat Joe Manchin of West Virginia finished demanding cuts, the bill included a staggering $700 billion in federal spending. Of that, $300 billion would be used to pay down the federal budget deficit, $69 billion for health care, and the biggest share—$369 billion—for clean energy. The Inflation Reduction Act was, more than anything, a global warming bill.

The reality was coal was dying in West Virginia, and groups like CCAN had blocked one of two proposed pipelines to ship fracked gas out of the state. So now Joe Manchin saw wind farms and battery-manufacturing plants—plus nuclear energy and less feasible carbon capture technology—as a big part of the future of his state's economy. Most analysts believed the IRA would actually trigger closer to $800 billion in federal climate spending over ten years and a stunning $1.7 trillion in private US investments.

Which led to that day on the White House lawn, September 13, 2022. The bill had passed Congress in August, and now we were there to celebrate, drenched in sunshine, more than a thousand activists, legislators, philanthropists, union leaders, and health care workers.

Before the Marine Corps band played and James Taylor sang a song and the president spoke, before the poems were read and a young union apprentice from Boston talked of a clean-energy world—before all that, I just walked around, enjoying this de facto family reunion of climate leaders after two decades of fight. We had made it. We hugged one another like marathon runners, from all across America, exhausted and happy after a race. There was Mary Anne Hitt of Shepherdstown, West Virginia, who, more than anyone, had helped launch the national movement to fight coal and stop mountaintop-removal mining. There was Saul Griffith of California, whose vision to "electrify everything" was now the rally cry of the climate movement as we transitioned to all-electric vehicles and homes powered by electric appliances run by wind and solar. There was Virginia state delegate Alfonso Lopez, founder of the clean-energy caucus in Richmond. And delegate Dana Stein of Maryland, a climate hawk among climate hawks in Annapolis and the friendliest man I've ever met. And there by my side was my superstar colleague at CCAN, Jamie DeMarco, our Capitol Hill lobbyist for an organization that now included twenty-four staffers spread across the region.

Afterward, I remember a feeling of floating as I walked from the White House to the nearby neighborhood of Dupont

Circle, where my family was waiting for me. Beth, my wife, had come down from Takoma Park. Sasha had walked over from his work at the US Department of Energy, looking all grown up at the age of twenty-five. My dad had flown up from Georgia. We celebrated at an outdoor restaurant on that most perfect DC evening on that most perfect DC day.

IN A PERFECT world, I would have stayed happy after the White House event. I wanted to stay happy. But soon, my dad flew back home and Sasha returned to his busy DC life and Beth left most mornings for her job as a local interior designer. I took a week off, the recovering marathoner, and mostly sat in my backyard feeling increasingly confused.

I don't know what I expected, but the news from around the world didn't change just because we passed a bill. Thirty million people were displaced at that moment in Pakistan from record flooding in the summer of 2022. The twenty-year drought in the American Southwest was still erasing Lake Meade and threatening to shut down the Hoover Dam.

Denial is a powerful thing. I didn't deny what was happening at home along Willow Avenue and across the neighborhood. I had seen the dying trees and the chain saws and the cranes long before September 2022. I had consoled my neighbors for their losses. What I didn't see was just how far behind we were in actually solving the core problem. I was totally focused on the smaller victories of state legislation in the run-up to a historic federal fight. That fight had happened. We had won—and thank god. We were doomed without a statutory national plan to transition off fossil fuels.

But it wasn't enough. Whether I knew it the whole time and wouldn't admit it—it was clear now: we weren't "avoiding the worst impacts of climate change." We had waited too long. There was too much carbon "banked" in the atmosphere, and it would take another twenty to thirty years to fully implement the clean-energy changes now on the books. Meanwhile, the very worst impacts were already arriving in many countries around the world. And in my region, we had already experienced the "land hurricane" derecho storm of 2012 and the atmospheric river of 2018, and there was ongoing, almost daily flooding in Virginia Beach and in downtown Annapolis.

Dr. James Hansen, the US dean of climate scientists and formally of the NASA Goddard Institute for Space Studies, had been saying for years that UN scientists and others were underestimating the speed of climate change. Coming out of my tunnel vision now, I believed him. All I had to do was look up and down my neighborhood and see all the missing trees.

And not just *see* them. That September of 2022, after returning from the White House, as I recuperated with books and naps in my backyard, the seasons began to change. The September wind blew. That melancholy sound, softer and hollowed out, returned to a neighborhood of missing trees.

Part II

ONE YEAR ON WILLOW AVENUE: 2023

3

The Undertaker

January–February 2023

Where do all the trees on Willow Avenue go when they die?

That autumn of 2022, sitting in my backyard, wondering what was next for the climate movement, I suddenly had a more parochial question on my mind: Where do the sawed-up trees actually *go*? I had asked the question several times of local tree-care companies, but the answers were . . . unconvincing.

I meant the question literally, not metaphysically, although if there's a heaven for any creatures on this earth, surely it's for trees. Beings that can live a century or more, offering habitat and comfort to all, asking little in return, uncomplaining—they don't deserve heaven? You look at the Miller Tree before she died—arms outstretched in graceful pose, weathered bark like artwork sketches, a canopy swaying in the wind—and you don't see a soul?

But where do all these wooden beings wind up when they are cut down and taken away?

That question came to me as I worried about my own big oak in the backyard. It was around seventy years old, a pin oak of the red oak family, with lovely seven-point leaves and a regal canopy. But it was also showing signs of illness there behind the house. After all the mortality on the block, this tree, *Quercus palustris*, was suddenly the second-oldest oak on the street. And it had been dropping limbs lately, something we'd never seen before. Coffee in hand, I looked up one morning to see a sizable widow-maker branch. It had broken off from the upper canopy and was hanging precariously in a tangle of lower branches, ready to fall and kill someone.

A tree service came and removed the branch. Workers sprinkled sea kelp on the ground below the crown to biostimulate the roots. Now, like so many Takoma Parkers, I was on "the watch." I thought of my dad's stories of parents in the 1940s and '50s, worried their children would be next to get polio. Or how I worried about *him*, my dad, in the 2020s, fearful he would get COVID in his late seventies. Now I fretted about my tree, its body and soul. Would it be the next to go?

I wasn't feeling great myself that fall. In this same backyard, in 2008, I had felt a slight tug under my shirt after a weekend of raking leaves. I looked down to see a half-submerged tick in my skin, hind legs still moving behind a buried head. For six years afterward, I was misdiagnosed. When I finally learned I had Lyme disease, an illness spreading throughout my neighborhood as milder winters fail to kill off the ticks, it was too late for a cure. Antibiotics would help, but they couldn't kill off all the spirochete bacteria that had now burrowed deep into every part of my body.

Today, I have what is called *chronic Lyme* or *persistent Lyme*. Tests show I've been infected more than once by tick bites in the last fifteen years. Now my weakened immune system periodically crashes and my body becomes inflamed. I take antibiotics, I recover for a while, and then I crash again. In the fall of 2022, I had a big crash. The week of rest I had planned after the White House ceremony turned into two. I lay on a backyard couch and wondered—for myself, for my tree, for my neighborhood, for my planet—*What now?*

One thing I knew for sure: when you get lost deep in the woods, when the trail peters out or you take a wrong turn, you're supposed to STOP. It's a concept conveyed early in the Boy Scouts' wilderness survival class, a class I taught for years when Sasha was a member of Troop 33. That troop had been meeting in the basement of the Takoma Park Presbyterian Church—my church—for one hundred years. My class met in the basement there and in my backyard, and then we camped in West Virginia.

Stop. Think. Observe. Plan. When you realize you're lost in the wilderness, you need to literally *stop*. Don't keep going blindly forward. Then you *think*. What actually happened? How did you get lost? Then you *observe*. What evidence is there for a possible route back to safety? Then you choose the route and *plan* your way out.

I had definitely taken the first step that fall. I was shut down, stopped. Normally, I can hide my Lyme disease from other people. I get through the workweek by distracting myself—my mind and my body—with the constant tasks of office life and intense field activism. I appear busy and okay. But then I crash on weekends. By midafternoon on many Saturdays, I've grown quiet at home, and my wife looks at me and says, "You don't feel well, do you?" My muscles are stiff, my body feels fluey, my brain in a fog. Then Monday comes, I rally, and the cycle repeats.

But the fall of 2022 was like crashing after a twenty-year workweek. I had been campaigning full tilt for so long. When I paused, the Lyme symptoms flooded my body as badly as ever. Dreading more antibiotics, I turned to acupuncture instead that fall and a big load of herbal medicines. We were sick together, my tree and I, climate change our common threat, getting new treatments in tandem.

Think. I did a lot of that, too. I reflected on twenty years of victories and defeats. I remembered Bill McKibben's famous quote, "In the fight against global warming, winning slowly is the same as losing."

Observe. That fall, I lined my bedroom and back porch with books on how we can become unlost in the climate movement. Each author pointed to a different path, beyond just clean energy, that could save us. I read about sequestering carbon through regenerative agriculture. I read about ships that could brighten ocean clouds with a spray of saltwater. I read about giant machines that could filter and capture carbon from thin air.

But a plan? None of these ideas was a silver bullet. There was no single plan to follow. Some were obviously doable but too expensive. Others were too early in development. Still others too dependent on unlikely human behavior change. But some mix of these ideas to capture carbon and reflect sunlight back into space would potentially be needed.

The one factor unifying all these ideas was this: planetary overshoot. We humans were about to overshoot the presumed safe level of warming for the planet—1.5 degrees

Celsius (2.7 degrees Fahrenheit) above preindustrial levels. To go past that temperature rise meant the Greenland ice sheet would surely melt disastrously. Mega-droughts would become common in the world's breadbaskets. With the planet already at 1.2 degrees Celsius entering 2023 and moving higher, there was no realistic way to avoid 1.5 degrees—or even 2 degrees—without extraordinary new interventions.

We were definitely lost. We were in a new wilderness.

Eventually that fall, it became too chilly to sit outside and read in the yard or lounge on the back porch. I gathered my books and my many notes and retreated inside. The backyard pin oak, meanwhile, throughout November and December, continued to drop branches. Limbs as long as I was tall kept detaching from the upper canopy. They tumbled down with the acorns and the autumn leaves that gradually covered the whole backyard, just ahead of winter.

WHEN IT CAME, the planet's hottest year on record began on Willow Avenue with trees trying to make babies way too early. As temperatures soared into the sixties in January 2023—and stayed well above average for most of the month here—the red maples on my block thought it was early March. Their buds began to swell by late January, and a few branches actually started to flower by the end of the month. Within a week, delicate puffs of tiny red petals, ready for pollination, emerged up and down the leafless gray branches.

Junipers and other evergreen shrubs began flowering and cranking out pollen of their own that winter, way ahead of schedule. On a walk in early February, I brushed my hand against the cones of an Oriental arborvitae and watched, stunned, as a cloud of pollen exploded from the shrub.

The planet as a whole began 2023 with the lowest January sea ice ever recorded in Antarctica and with serious drought covering big parts of five continents. No one knew it was going to be the earth's warmest year, of course, surpassing 2016 by a stunning margin. But a heat-enhancing El Niño system was forming in the Pacific Ocean—and before the year was out, there would be devastating rains in Libya, a wildfire in Maui killing one hundred people, and ocean water so warm in the North Atlantic it drove native fish into the Arctic to survive.

But on Willow Avenue, it began with out-of-sync trees. January and February brought the third-warmest start to winter on record in the DC region. That was followed, weirdly, by one of the chilliest late springs in memory and then by wildfire smoke from Canada that contaminated our summer air. That, in turn, was followed by a sharp drought in the fall and an end-of-year deluge of rain in December.

But first came the winter warmth. It did more than disrupt the trees' reproductive schedules. It scrambled the core metabolism of many trees, including oaks—creating life-threatening conditions for some. In the past few years, many of us in Takoma Park had noticed that, among a growing number of trees that survived the soaked ground of 2018 and the subsequent attack of ambrosia beetles, there was a new and strange syndrome emerging characterized by leafless "fingertips." At the top of the canopies, the branches were bare in summer—no leaves at all—for their last three or four feet of growth. This created the disturbing look of bony fingers reaching up through the canopy.

Several arborists told me the same thing: "We think the erratic winters are contributing to this." Like most trees, oaks need to sleep in winter. They go dormant. To keep their branches from freezing in the cold weather, the trees convert the starch in their branches to an extra-sugary sap. It serves as an antifreeze. Plus the branches expel water from their living cells, reducing the chance of freezing. But when winter turns prematurely warm in mid-season, as now happens so frequently, the trees start to wake up and pump water into their extremities, thinking spring is arriving and it's almost time to make leaves again. But soon, as the roller-coaster weather turns back to freezing, the branches are suddenly vulnerable and less protected, their defenses diluted. Cells in the branch tips start to freeze and die. If it continues for too long and the tree loses 30 percent or more of its spring foliage, it will likely die. Was this, I wondered, what was happening to my pin oak?

The winter of 2023 on Willow Avenue was as erratic as they come, cruelly tricking and stressing and harming trees. No wonder even as the beetle infestation in my neighborhood had slowed for now, tree deaths in the city were still running at twice the recent historic average.

Which prompted, again, the question: When the end of the line finally comes, when the trees die here, where do their bodies actually go?

"I THINK I know the answer," Ning Zeng told me when I posed the question to him. "I think I know where the dead trees go."

Ning is my neighbor. He is also a runner. When he runs, he thinks about lesson plans for his students and about the anaerobic decay experiment he's conducting in his basement laboratory at home. Ning is a professor of climate science at the University of Maryland.

One day, Ning ran past the public works compound for the city of Takoma Park. It's tucked out of view on a hill on Oswego Avenue. There, Ning was shocked to see a mound of dead trees. Dozens of sawed-up trunks were haphazardly piled on top of one another, the bleached bones of trees that once protected and shaded this neighborhood.

The tree companies that do the cutting around here like to tell you the trees are mostly "repurposed" into mulch for gardeners or cut into boards by small local mills. But Ning was discovering otherwise.

I knew none of this when I invited Ning to my house one Sunday morning in early February 2023. All I knew was that he, too, had recently lost a big white oak to beetles in his yard. He lives in Silver Spring, Maryland, right on the border with Takoma Park, barely a mile from my house.

"It was an extraordinary tree," Ning said of the beauty in his backyard. He had raised two kids under its branches. "It took me and my children to wrap our arms around it. I put a trampoline under the branches for them to play."

Ning speaks with the noticeable accent of someone who moved to the US at age twenty-two—from China's Sichuan Province. He came for graduate school and stayed. He's fifty-seven now.

Like most people in the DC region, even Ning—the climate professor—didn't realize his lost tree was part of a specific pattern. I'd spent much of the fall of 2022 learning all I could about the jet stream and the phytophthora mold and the reproductive skills of ambrosia beetles. I'd interviewed local tree specialists and leading entomologists as far away as the University of Wisconsin, trying to piece together the story. Then in December, the highly respected *Bay Journal* ran a story summarizing the alarm of forest managers across the Chesapeake region over rampant oak tree deaths. Those experts stressed the "unsurpassed importance" of oaks to urban and rural forests, with acorns feeding more than one hundred species of animals and the oak tree leaves serving as a primary food for caterpillars that in turn feed migratory birds. And, according to the *Journal* article, "Across the board, experts ... cite more intense weather from climate change as a major cause in the seemingly sudden fatalities."

Having done some homework, I now wanted to give Ning, a contributing scientist with the UN's famed Intergovernmental Panel on Climate Change, a tour of what that region-wide trend had done to the streets around my house.

Ning is a talker—in a good way. In five minutes, he'll jump from the latest climate research on hurricanes to drought patterns in the American West to how new satellites are monitoring forest fires worldwide. All the while, he'll hold your attention with a storyteller's tone and a pair of kind, dark eyes that look right into your own.

But the moment we left my house and crossed the street and I showed him the remains of the Miller Tree, Ning grew silent. There it sat, a darkened trunk, four feet in diameter, behind the porch. It had been nearly two years since the tree's death.

"Oh my," Ning finally said. "Oh my god." He walked slowly around the massive stump, palming the flat surface where more than 150 growth rings were still visible. He ran his finger along the decomposing outer bark, now crowded with blueish-gray lichen. He was silent again.

"There are more," I said.

We walked past the Kurtz-Greenberger house, where a towering white oak once stood in the backyard. With the sweep of my hand, I pointed out to Ning the large gap in the canopy that now runs nearly the length of the block behind the houses to the northwest.

We turned next onto Tulip Avenue and another "headstone" a block away. It sat kitty-corner from my church at Tulip and Maple. The trunk was almost as big as the Miller Tree. On one side of the trunk, you could see a "lightning scar" from where the tree had absorbed a massive bolt of electricity in 1974—and lived. But the 2018 rains and other weather extremes were too much. "A foreigner in its own land," as Chris Larkin would say.

Ning still said little as we kept walking. He wore horn-rimmed glasses on a face both boyish and serious, framed by black hair with tiny flecks of gray.

We passed three tree stumps in a single yard on Maple Avenue. Then came the amputated trees—not yet dead but dying. Homeowners would lop off giant dead branches to try to stem the decline, giving some of the trees the sad look of delicate sculptures from antiquity, armless and incomplete.

"No wonder," Ning started saying. "No wonder."

He elaborated: "I knew oak trees were in decline around here, but I never realized the whole scale. No wonder so many cut-up trees are piling up in the places where I go."

As a climate scientist, Ning is obsessed with the problem of planetary overshoot. We humans, again, were blowing past the safety zone for global temperature and carbon in the atmosphere. He's also obsessed with trees. He thinks he's figured out how to use trees in a surprising way to help stabilize the climate and bring us back to a safe earth. And every day, he sees greater urgency for his idea.

Case in point: Back at my house, I showed him a map of Takoma Park that the city arborist had recently given me. The boundaries of this small town had been overlain with over a thousand small dots. Each dot represented the site where a damaged or diseased adult tree had been cut down along city curbs or on private property with a permit during a recent twenty-four-month period.

What strikes you most is how evenly spread out the dots are on the map. It's as if a computer had perfectly randomized the dots. Until recently, one city official told me, those fatalities would have been more bunched up—a thunderstorm microburst here or there knocking down clumps of trees on a block or, as with the famous 2012 derecho, a concentrated string of downed trees along Sligo Creek as the hurricane-like wind from that storm was funneled into the valley. But this map was something else. Trees were dying evenly on almost every block, a perfect distribution. "It's clearly systemic," said Daryl Braithwaite, the city's director of public works, when we discussed the map earlier. "That doesn't look like weather. It looks like a climate." My small town, on this small map, presented evidence of a new paradigm of loss.

That's when, back at my house, sitting at the dining room table, Ning said for the twentieth time that morning, "I want to bury them. I want to put all these trees in the ground."

CLIMATE SCIENTISTS CALL them *negative emissions*. When you hop into your car—and it's powered by gas—you're emitting CO_2 all the way to the grocery store. That CO_2 migrates to the atmosphere and traps heat for a very long time. If we could figure out a way to suck that same CO_2 out of the air and store it safely somewhere on Earth, it would be a negative emission, a kind of reverse pollution. We erase our sin.

The problem is it's not that easy to do. For the past 250 years, since humans first combusted fossil fuels to begin powering the Industrial Revolution, we've been lucky. The world's oceans and forests, through complex means, have absorbed more than half of the excess CO_2. But now the oceans have slowed their absorption of carbon and grown acidified from what they've absorbed already, and many of our largest forests are being logged, are drying out, and are burning down.

So now what? As always, the first step is to stop making things worse. Stop burning fossil fuels.

Simultaneously, we must find a way—on a massive scale—to "draw down" the excess CO_2 and store it. Global temperature and CO_2 correlate closely. Bring down carbon in the atmosphere and temperature will follow. It's not a chicken-and-egg problem. CO_2 comes first here. Temperature second.

So the goal, scientists say, is 350 parts per million (ppm) CO_2 in the sky. At that level, we can maintain the stable temperature and climate that human civilization has thrived in for the past ten thousand years. But here's the problem: we're already at 420 ppm carbon in the atmosphere right now, and it's rising fast. We're *already* in unsafe territory, wandering in a wilderness, off the beaten path. And as UN secretary-general António Guterres says, there's a cliff up ahead. It's approaching fast. Only he's wrong. With CO_2, we've actually already gone *over* the cliff. We're hovering in midair on the other side. We're in overshoot.

The question now is: Can we—in midair—weave together some kind of rope to pull ourselves back up to the ledge once we've finally stopped descending on the other side?

Actually, we might need two ropes. One is the negative emissions lifeline, sucking CO_2 out of the atmosphere. The second rope, in case the first one is not strong enough, is the SRM rope, solar radiation modification (i.e., reflecting sunlight away from the planet—more on that later).

A majority of scientists working on the overshoot problem are focused on the first rope: removing CO_2. The work is vital.

Regenerative agriculture—also known as *carbon farming*—is probably the most promising potential carbon-reduction technique. What if we flipped modern agriculture from a climate *problem* (CO_2 spewing from tilled soils, methane released from rice paddies) into a climate *solution*? The ideas here are legion, and theoretically, most of the world's excess carbon could one day be locked into soils. How? We could pay farmers worldwide to use no-till farming methods and start planting cover crops in fallow periods. We could enhance soils with "biochar," a charcoal-like material that bolsters production. But questions abound. What will all of this cost, and who will pay? And how, in ten years or less, do we overcome the inertia of Big Ag and the regional cultural resistance of small farmers worldwide?

Another CO_2 idea is this: make rocks work for us. All over the world, all the time, rocks undergo natural "weatherization" when they are exposed to air and water. As one study explains: "When silicate or bicarbonate minerals in the [rocks] dissolve in rain water, carbon dioxide is drawn from the atmosphere into the solution to form bicarbonate ions." These ions eventually flow to the ocean for more or less permanent storage. Currently, one out of every forty carbon dioxide molecules in the atmosphere is absorbed by rocks. But if we mined and crushed certain rocks like basalt and spread the dust on farm fields as fertilizer, the number could grow severalfold higher.

The problem again is the potential cost and the environmental impacts from vast mining operations and the global transport of rock dust.

Still another much-discussed technique is using those giant machines to capture CO_2. Freestanding structures would be built around the world that filter air and capture carbon through chemical reactions, converting it to liquid or solid. Afterward, heat is applied to re-gasify the carbon so it can be transported as CO_2 to storage reservoirs underground. If it sounds complicated, it is—*and* energy-intensive and expensive. This "direct air capture" technology is still very much in the development and prototype phase.

But why use machines to capture CO_2 when you can use plants? Photosynthesis is the most powerful and common form of carbon capture. Yet exploiting it as a climate response is not a straightforward task. One idea, given too much credence in my view by the Intergovernmental Panel on Climate Change, is something called *bio-energy with carbon capture and storage*. This involves farming fast-growing grasses or other plants, then burning them to create electricity and capturing the resultant CO_2 and storing it underground. Among the many problems is scalability. To have a major impact, as much as 20 percent of all the world's agricultural land—an area the size of Australia—would have to be converted to energy "feedstock crops" to supply power plants.

Most of these carbon-sequestration techniques—and many others—have some kernel of promise. But none—solely or in combination—is even remotely on track to getting us the reductions needed by our warming planet: ten billion tons per year by 2060, according to the IPCC. There's no rope here—yet—to pull us back up to the ledge. This infancy of technology and scale is terrifying since the IPCC—the ultimate scientific advisor to world leaders—is now betting the farm on negative emissions. The most recent IPCC assessment reports are clear: without these carbon sequestration techniques working as desired, clean energy alone has no real chance of bringing CO_2 and temperatures into the safe range.

All the more reason to hope somebody, somewhere, comes up with more ideas.

Somebody like Ning Zeng.

IT WAS AN insanely warm February day the first time I saw that chaotic lot of chopped-up trees outside of Baltimore, their trunks piled high into small mountains. Ning took me there.

We had toured my block the week before. Now, on February 15, he was showing me a sample of where the region's dead trees wind up.

I woke to daffodils in full bloom on Willow Avenue that morning. I had never seen so many of these yellow flowers this early here, starting in late January. Purple crocuses and white snowdrops were blooming, too, all over Takoma Park, arriving as early as *mid*-January. Another first, as far as I knew.

Before leaving for Ning's house and the tour of unburied trees, I was tempted to plant a few snow peas in my vegetable garden. You traditionally plant these peas around St. Patrick's Day here. But that advice had been dismissed long ago as passé pre-warming wisdom. Why not mid-February? Feeling lucky?

The red maple trees on Willow Avenue were now in full flower, as were star magnolias all across the city, nearly a month early. On February 8, according to *The Washington Post*, the tree pollen index set a record in DC for the date.

There were 487 grains of pollen for every cubic meter of air. Sasha was taking his spring allergy medicine most mornings in early February, and the NFL football season wasn't officially over yet. The story was the same across most of the East Coast, with New York and Philadelphia off to their warmest starts ever.

My friend Cam, a climate activist, says, "Trying to solve global warming while global warming is crashing down on your head is like trying to learn calculus inside a burning building."

It sure feels that way some days.

By the time I got to Ning's house that morning, he was standing on the curb, ready to go. He'd already spent an hour in his basement lab, checking again the estimated tons of tree corpses he wanted to collect. "I want to purchase fifteen thousand tons," he said.

We jumped in Ning's 2002 Toyota Corolla, with stick shift. "I'm on a waiting list for a solar-powered car," he said. "It has three wheels and it'll get me to my campus office totally on sunshine. But too many people want it. Delivery in 2024."

The hour-long drive to Baltimore gave Ning time to tell me more of his life story. It was a tale of trees and climate change. Here it is:

There were no forests near the small Sichuan village where Ning grew up in central China. Rice paddies and small fields dominated the landscape in all directions. There was one tree, however, a huge red bean tree (*Ormosia hosiei*). It grew right outside Ning's grandparents' home. It took five kids to form a ring around it. The species is planted beside houses in this part of China because, Ning said, you experience a feeling of love under its branches. Your chest fills up with it. The leaves are heart-shaped and flowers pinkish red in spring.

That tree canopy helped nurture the Zeng family until 1965, the year the violent Cultural Revolution began. Ning was born that year, right before his father was taken to a Communist reeducation camp and tortured. Sinking into poverty, Ning's mother gave him away to be raised by friends for a year. After his father survived and the family was reunited, it took years of steady love from that red bean tree to heal those many family wounds.

Ning, as a child, always wanted to be a scientist after reading about Albert Einstein. In his twenties, climate science became his passion. When he attended college at the University of Science and Technology near Shanghai—the "MIT of China"—there were no forests nearby for him to visit. And none in the desert setting of the University of Arizona, where he went to graduate school. And no forests amid the sprawl of Los Angeles for his postdoctoral studies at UCLA. Only when his studies took him to the NASA Goddard Space Flight Center in Greenbelt, Maryland, did he finally get to live and work near a forest.

"My first week at Goddard, I walked out of Building 33 and there it was, a big, deep place of trees that seemed to go on forever."

Soon he was wandering into the forest, without a trail, taking it all in—and he discovered big fallen oaks and other trees, decomposing on the forest floor. He had studied the global carbon cycle for years by then. What if, he wondered, instead of decaying and releasing tons of CO_2 back into the atmosphere, some of these trees were buried? He imagined an oxygen-free setting of some kind, underground carbon storage for a long, long time. A "wood vault," he called it. Would it help fight climate change? What if you buried thousands of trees? Or millions? Or billions?

Over the years, he played with the numbers, conducted a few small experiments, and stumbled on articles about wooden Viking ships buried for a thousand years and barely decayed. Then, in 2018, the IPCC for the first time made clear that negative emissions on a large scale were needed to avoid climate calamity. So Ning refined his numbers one more time and joined 1,133 other teams of scientists and tinkerers in applying for an XPRIZE for carbon removal from the entrepreneur Elon Musk. It was worth $50 million.

On Earth Day 2022, he was notified he was a finalist.

"Look it up on Google Maps," Ning said as we got closer to Camp Small in Baltimore. "Some crazy photos pop up. Who ever heard of such a place?"

Suddenly, on my phone, there were incomprehensible walls of gray wood lining a dirt road. There were photos of cars loading up with firewood, yet barely making a dent in the mounds all around.

But the actual place, when we arrived, barely resembled the outdated online photos. Those small mounds had grown into low mountain ranges of wood. So many more trees had arrived here in recent years. So many.

Ning handed me his phone and asked me to film as we entered the five-acre compound, marked only by a handwritten sign: "Camp Small." Suddenly, the road was like a valley floor with high woodpile ridges on either side. We were surrounded by sawed-up trees of every size and shape, with two-hundred-year-old trunks stacked together with younger trunks and a kind of messy caulking of smaller branches in between. On we drove, along the valley floor, millions of pieces of wood around us. Off to the right veered another road, into another valley, which branched off again into still other valleys.

I saw crows overhead, banking in the sky. From up there, I thought, we must look quite small in Ning's compact Corolla, rambling across this maze of wooden bones, fragments of former life piled twenty-five feet high in places, a junkyard purgatory of walnut trees, pines, ash, tulips, maples, and, overwhelmingly, oaks.

This is where the dead trees go when they die in Baltimore—at least on city property. On parkland, at schools, along street rights-of-way—they wind up here, the municipality's tree dump. Starting in 2016, as part of Baltimore's new zero-waste plan, dead trees from city land were diverted away from local landfills and trash incinerators. So Camp Small was dedicated just north of downtown as a transfer station. Mulch companies and lumber businesses could come and buy their fill. Anything left over could be milled—on location—into boards for city park benches and flower boxes. At least that was the idea.

The problem was that supply was outstripping demand—here and throughout the region. There were more trees dying than wood buyers could consume. The market for wood chips and mulch was absolutely saturated among landscapers. After that day in Takoma Park, when Ning passed a mass of dead trunks piled up on city government property, he began calling around to tree-cutting services. "You want some trees?" one company in Gaithersburg, Maryland, told him. "We've got a thousand tons right outside our office. Come 'n' get 'em. No charge."

Then Ning found Camp Small, the mother lode of all collection sites.

The day we visited, winding through those stacked-up tree bones to the center of the yard, we met the yardmaster, Shaun Preston. He was a cliché of flannel shirt, full beard, and baseball cap—an urban lumberjack working ten minutes from downtown Baltimore.

"We've seen a huge increase in trees here since I started in 2016," Shaun told me as another city truck arrived with more trees. It beeped loudly as it backed up to a fresh mound. A type of backhoe with mechanical jaws noisily lifted the trees onto a new pile as Shaun kept talking, having to raise his voice.

"Fifteen percent," he said.

"What?" I asked amid the noise.

"Fifteen percent. That's the amount of trees I can sell as mulch or lumber. The rest, all of what you see here, I gotta figure out how to move them off the lot. I've got to make space for more trees coming."

Shaun wasn't quite sure why, in recent years, so many more trees were arriving at the yard. "A lot of beetles, right?" he said.

"That's part of it," I said. Then I asked, "Are you seeing a really big increase in white oaks?"

"Yeah. How did you know that?"

I explained that white oaks, the state tree, were particularly susceptible to the phytophthora mold. The rain, the mold, the beetles. Climate change.

"Are you kidding me?" Shaun said. "That's why? Do you know how many dead oaks are still standing on Baltimore streets? They haven't even all arrived here yet. Climate? Where the hell am I gonna put them?" Then he added, "I'm glad Ning called."

Ning was running around the compound, taking photos and carrying a yardstick for various measurements. It was 11:00 a.m. and getting crazy warm—just one day after Valentine's. Ning stripped down to a T-shirt that said, FEEL THE HEAT! He wasn't sure where he got it.

On one end of the yard was a virtual mountain of mulch and wood chips, rising above the peaks and valleys of dead logs all around. Ning and I half walked and half climbed to the top, wood chips pouring into our shoes.

Once at the summit, Ning surveyed the sprawling yard and described his vision. As a pilot project for the world to see, using big timber-hauling trucks, he would transport the first five thousand tons of these trees, one-fifth of the lot, to a farm two counties away from here. That land had perfect soil—a solid variety of clay. A backhoe would remove the topsoil, then dig a long, rectangular trench in the underlying clay. It would be about an acre in size and fifteen feet deep. The trees from Camp Small, from Baltimore's streets, would then be lowered into the trench, filling it. They would be stacked on top of one another in sections, five rows per section, about ten trees per row.

The whole thing would then be covered with the excavated soil, creating a long, low mound, everything covered with the airtight clay. Then the topsoil would be returned, then grass planted, then you would have room for a solar farm on top or a park promenade or just leave it as grass for grazing animals. It would be held, undisturbed, as a conservation easement, the land donated by the landowner. Even after factoring in the CO_2 emitted during tree transport and burial, a net 95 percent of the trees' carbon would be preserved. Ning predicts the trees will be safe here, in the near-total absence of oxygen, for a thousand years or more. It is borderline geologic storage, he says. Others call it "reverse coal."

Of course, if this same wood were high-enough quality, you could use some of it instead for furniture or building materials, and it would also be a kind of sequestration. Only it usually winds up in a landfill, Ning said, within twenty-five to seventy-five years, if not sooner. There, in temperatures as high as 170 degrees Fahrenheit, thanks to the presence of food scraps and other organic material, the wood decomposes pretty fast, sending CO_2 and methane leaking into the sky. "A bioreactor" is what Ning calls landfills.

Better to bury it in tightly controlled structures, smothered in clay. Sophisticated sensors for methane and carbon dioxide will be stationed inside the mounds and on top of them to verify safe storage.

"That's the plan," Ning said as we stood atop that towering mound of mulch in Baltimore. It was almost noon, and we were getting sweaty in the hot February sun, the temperature rising to seventy-one degrees. It didn't help that under our feet, plumes of steam—like smoke—were rising up from the woody mulch decomposing below us. This pile was a de facto compost heap, venting CO_2 into the atmosphere as we spoke.

Just then, a fox appeared on the dirt road below us. It paused, stared at us, then disappeared into the craggy maze of wooden peaks and valleys. Another logging truck appeared soon after that, full of new trunks and limbs to be dumped into this rambling field of castoffs. Before the week was out, dozens more trucks would arrive, packing this place to bursting, compounding the city of Baltimore's ever-growing headache of just what to do with these ever-growing stacks of dead trees rising toward the sky.

THERE ARE APPROXIMATELY three trillion living trees on planet Earth today in urban, rural, and wilderness settings. We need to keep them healthy and alive. They sequester nearly a third of the carbon emissions we pump into the atmosphere each year. And scientists say there is enough suitable land on Earth to plant an additional one trillion trees.

So to weathered rocks and regenerative farming and all the rest—add these two fundamental ways to store carbon: planting trees and preserving forests. The data is strong on this, and governments worldwide are mostly supportive.

But it takes decades for newly planted trees to reach a size able to draw down substantial carbon. It's *not* a quick fix. And it will take an area the size of China to plant those trillion trees, likely infringing on the lifestyles and sovereignty of human communities in many areas, especially in the global south. Plus, again, we need those new trees to stay *alive*—not a guarantee in a warming world.

Same with our existing forests. We need them to survive in a shifting climate. Many are not. Already in California, a million acres of forests specifically set aside as carbon offsets to sequester CO_2 have burned in recent wildfires.

Ning believes planting and protecting trees worldwide is a top priority. But he also wants to be there when they die, stopping their release of carbon into the atmosphere wherever possible. Like the trees at Camp Small, he wants to bury millions of Western conifers now dead or dying from bark beetle attacks linked to climate. The idea is to bury the trees close to where they fall, in earthen vaults, like the one planned for the Baltimore trees. The same for trees killed by increasingly strong hurricanes along the Gulf of Mexico coastlines. It's impossible to bury all the downed trees. But bury as many as you can.

More immediately, after a century of fire suppression out West, the US Forest Service wants to remove six hundred million tons of dead trees from the forest floor in twelve states over the next ten years. The service has money from the Inflation Reduction Act to do so. Inconceivably, officials there want to burn the trees in giant waste pits. Ning wants to bury them instead.

But to really make a difference worldwide, Ning says, you would need to manage nearly half the healthy forests across the planet so that a portion of those trees are removed upon natural death and buried, leaving enough to fall and decay and maintain healthy soils. Altogether, these methods could store one to two billion tons of carbon per year, up to a fifth of what the IPCC says is needed as negative emissions by 2060.

The sequestration merits of buried trees are undeniable. Beyond Viking ship anecdotes, Ning worked with colleagues at McGill University in Montreal to bury thirty-five tons of waste trees east of the city in 2013. Eight years later in 2021, they checked and found samples from the site in near-pristine condition. Perhaps more telling, in the process of digging, they accidentally found an ancient cedar tree lodged in adjacent soil. It was still in log form and had bark attached.

A local lab ran the tests. The tree was 3,775 years old.

* * *

"I'M NOT A religious person," Ning said. "But I'm kind of spiritual—for a scientist. That's what my wife says."

He chuckled.

We had just climbed down from the mountaintop of mulched wood. We were walking along another dirt road, more ridges of dead trees on either side. I was asking Ning how he was feeling, deep down, about . . . well, everything—the loss of his tree back home, the mortality across the region, the appalling evidence of climate chaos right here in this overwhelmed yard.

He tried to maintain his scientist role at first. Molecules and engineering equations were on his mind, he said. How could he show the world that wood's carbon, hydrogen, and oxygen were stable underground?

"No," I said. "How do you *feel*?"

More than anything in the world, I knew, Ning wanted to give a proper burial to the dead and dying trees of Baltimore and Takoma Park and across the globe. But the proposition was more than a math equation for him.

"I feel love," Ning said. He jumped up on a big log right next to the road and gestured to the wood all around us. "I feel love for these trees. All of them." He said it was like being around that red bean tree at his grandparents' home in China. "Love is filling my heart right now," he said. He touched his chest.

"And I feel urgency. I feel time is running out and we can't let these trees slip through our fingers. They stored carbon all their lives and helped the planet fight climate change until they died. Now, if we bury them the right way, they can help keep future trees alive."

In Ning, great strands of mercy and science came together. But he was way ahead of society in appreciating the value of dead trees. The City of Baltimore was selling him fifteen thousand tons for $150. That's it. A penny per ton.

Morning had turned to afternoon in this woodlot. A mouse scurried in the shadows. The yardmaster had told us these trees—even dead—gave shelter and life to every imaginable critter: mice, marmots, black snakes, feral cats, foxes, and, overhead, red-tailed hawks and crows.

As we prepared to leave, Ning, the scientist, the tree undertaker, said he wanted to have a ceremony when the last of these trees goes into the ground at the rural farm field planned for them.

"We'll read a poem," he said, "and people will speak. And then we'll cover up that last tree and have a long moment of silence."

4

Cherry Trees

March

Pink-blooming clouds of cherry trees come to Willow Avenue and Takoma Park in the month of March. The clouds float flawlessly a few feet off the ground, lining sidewalks and gardens, hanging on despite spring breezes that try, gently, to scatter their splendor.

Washington, DC, is famously obsessed with the puffy canopies of its cherry trees in spring. But the trees are equally popular in the neighborhoods and towns surrounding the capital, creating a region where cherry petals, when they finally fall, take the place of the snowflakes we rarely see anymore in March.

In those spectacular blue-sky days of early spring, the ornamental cherry trees—Yoshino, Kwanzan, Akebono, Usuzumi, Fugenzo, and others—put on a show so gorgeous they sow a dose of sadness with the joy. Each tree, with each flower, brings forth a thousand years of Japanese Buddhist insight: Life is beautiful. Life is fragile. Life is stunningly brief.

Life is also scrambled and chaotic on Willow Avenue and across this region—something I try to forget, momentarily, as I take in the mid-March bloom of two Yoshino trees at the end of my block. There near the DC line, a handful of five-petal flowers emerge at first, ahead of the peak, offering a white brilliance and soft fragrance that will soon dominate the senses of passersby. Later in spring comes the darker blossoms of a double weeping cherry tree on the street. The drooping branches cascade toward the ground with ten-petal flowers, a fountain in the middle of the block, planted years ago by Rohini Pande and Michael Gordy as a gift for their daughter, Hemakshi.

But even in this, our most beautiful time of year, there is no hiding the changing climate. Along the Tidal Basin in DC, where the main show of Yoshino blossoms occurs, tourists sidestep sections of flooded sidewalks at high tide. Entire park benches rest in several inches of water in some spots. West of the Jefferson Memorial, the stumps of cut-down Yoshino trees are visible while others struggle in the soggy soil of increasingly eroded banks thanks to sea-level rise and sinking land.

In 1912, when cherry trees were first planted around the Tidal Basin as a gift from the Japanese government, the world was at 301 parts per million (ppm) CO_2 in the atmosphere. The trees bloomed in early April then. Today, the world is well past 420 ppm, and the blossoms peak as early as mid-March some years.

Time. There's not much left to save our fading seasons. The cherry trees bloom so early now—here and worldwide—because the clean-energy revolution bloomed too late. And now, having spent time with Ning Zeng, the Maryland professor, I knew drawing CO_2 out of the air wasn't going to save us anytime soon.

The clean-energy part, to be sure, *was* racing toward full bloom nationwide. Since the passage of the federal Inflation Reduction Act the previous August, every day seemed to bring word of a new mega–battery factory for cars in Texas or a solar manufacturing plant in West Virginia or news that heat pumps—for the first time—were outselling gas furnaces for heating homes across America.

But there was no equivalent in Ning's world. No announcement of new factories to produce a million direct air capture machines for CO_2; no global treaty committing a trillion dollars to planting trees or paying farmers to transition to regenerative agriculture. Ning himself said, absent an XPRIZE award from Elon Musk in 2024 or the emergence of some other elusive investor, it would be hard to produce more than a few pilot projects for buried wood. His small company, a special benefit corporation called Carbon Lockdown, was running low on money to pay a team of contractors and engineers, and several were owed back pay already, he said.

One night on the phone, I told Ning it seemed like the world of negative emissions technology was today where solar technology was in the 1990s.

"No," he said abruptly, "you're wrong. It's where solar was in the *1970s*."

No wonder sixty of the world's top climate scientists, on February 27, 2023, issued a letter that instantly made history in the fight against climate change. Led by luminaries like James Hansen of Columbia University and Sarah Doherty of the University of Washington, the open letter declared it is "unlikely that [carbon dioxide removal] could be implemented rapidly enough or at sufficient scale to entirely avoid dangerous levels of climate warming in the near term."

After that bombshell statement, the scientists went on to recommend research and testing of solar radiation modification, a topic previously considered by many of these same scientists to be too controversial—or outright dangerous—to even study. Now they were calling for "rigorous, rapid" experimentation on how to reflect sunlight away from the planet, including how to disperse sulfur aerosols in the stratosphere.

But what really got my attention was *why* these scientists were proposing such an abrupt pivot. In the letter's very first paragraph, they wrote, "Natural systems are approaching thresholds for catastrophic changes with the potential to accelerate climate change impacts beyond humans' ability to adapt."

Beyond humans' ability to adapt. They weren't talking about periodic events like Hurricane Katrina in New Orleans or flooding in Pakistan. Those were one-off calamities contained to a specific region, followed by long and slow recoveries. Hansen et al. were warning about global and *systemic* failure. The failure of humans to adapt to recurring, violent climate patterns almost everywhere, all the time. A Katrina every year. A Pakistan that never stops flooding.

In that context, the financial bankruptcy of a city like Takoma Park, Maryland, would probably qualify as evidence of inability to adapt. While sea-level rise was killing cherry trees along the Tidal Basin seven miles to our south, the rapidly changing rain patterns were still killing many of our biggest oak trees on almost every block. That same water was flooding homes and structures like my Presbyterian church, one block from my house. Indeed, the disappearing trees and increased rain were now joining forces to accelerate water runoff everywhere, compounding the threat to human structures. And now my city—not in Louisiana, not in Pakistan—was starting to see the first outlines of a financial crisis looming ahead. The town was already shelling out hundreds of thousands of dollars for flooding it had never budgeted for. Any further increase in cascading water would tip those investments into the millions and, possibly—gasp—much, much more.

During that spring season of 2023, of course, the cherry trees across my neighborhood cared little about this approaching financial crisis. They continued their lavish blooming, earlier than normal per recent trends. The Yoshinos along Willow and Jackson and Holly Avenues were particularly showy that year. Those white flowers, tinged with the palest pink, were as fragile as watercolor brushstrokes, each offering again the ancient message: Time is short. Time is short.

"It came on so suddenly," Kurt Lawson said. "By the time we got there, the street was a lake, and water was just spilling down the walkway toward the preschool. We piled up sandbags as fast as we could. But it was leaking into the building before we knew it."

So began Kurt's telling of the locally famous flood of September 10, 2020. He and I were standing on the sidewalk outside our church, Takoma Park Presbyterian, on a decidedly non-stormy day in March 2023, a cloudless blue sky overhead. Two-plus years ago, on this sidewalk, is where the flood began.

Kurt was the chair of the church's property committee. In his sixties, with a handsome silver-white beard, he was dressed in khaki work pants, boots, and a flannel shirt marked with the white splashes of a recent church paint job.

First, some background. Over the past decade, Kurt said, the water problems had gotten worse and worse for the church. He told how the floor of the basement assembly room was now buckling and sinking, apparently from poorly compacted soil combined with what some church members believed was poor drainage from the increasingly intense rainfall in recent years. The slate roof of the sanctuary had also been leaking for a long time and had just been repaired at a fantastic cost. And now—now—during big storms, water was starting to enter the church's preschool doors, rushing past the down-sloping playground and overwhelming the French drain at the bottom of the hill. That drain had been working adequately since the 1960s but, in the past few years, was becoming insufficient for the new volumes.

"Then the September 10 storm came," Kurt said, "and it was bad, and we knew we couldn't let it happen again." The church boiler room flooded to ankle depth, and the assembly room floor was covered with water. No one had seen this before. But it was the preschool that caused the greatest concern. The teachers and attending youth, from ages eighteen months to twelve years, were evacuated away from the doors that day. Carpets and some furniture were damaged, but most of the water—a foot deep outside—was thankfully, narrowly, kept from fully entering the building. "But what happens next time?" Kurt asked. "How high will the water go?"

Kurt told me how the church leaders later went to city officials and said, "You've got to help us. We've got a real problem here." The Public Works Department promptly told the church that, due to water projects in multiple spots across the city, the municipality's stormwater management budget, all $700,000, was spent for the year.

"But you've got to help us," the church leaders repeated.

So the city came up with a novel solution. There was money in the sidewalk budget. What if, the city proposed, it busted up the perfectly good sidewalk along the church's Tulip Avenue side, removed the concrete, added multiple tons of soil to elevate the walkway up to fourteen inches, poured new concrete, and called it a day? Added bonus: the new sidewalk would be wheelchair accessible.

And that's what the city did. It built a de facto two-hundred-foot-long small *levee* to protect the church's preschool from flooding, finishing the work in 2022. Technically, the term is flood *berm*, and you have to know what you're looking at to understand that the oddly elevated sidewalk, with dirt sloping away on both sides, serves as a water barrier. No flood pipes were installed. No pumps. Just blunt-force soil injection because it was cheaper and would get the job done—for now.

By the time Kurt finished the story, there atop the new sidewalk berm, my head was spinning. I had known the story's outline beforehand, but the details now rekindled questions that many of us in the congregation had: How long can this go on? How will we know, as a church, as a neighborhood, when the changing rain patterns have gone so far that we can't adapt anymore? How will we know when financial and structural thresholds have been crossed for good? Are we there now?

What we knew for sure, as a congregation, was that the eighty-one solar panels on our church roof were not going to stop the next flood. Nor were projects like Ning's tree burial idea, at least not anytime soon. And it was clear to me, meanwhile, that what was happening here, in the middle of Takoma Park, was probably a pretty good proxy for city streets and stressed-out societies everywhere.

The ecological domino effect was perhaps the most disturbing part of the 2020 flood here. The record water barreling toward the church that day was runoff mostly from the 7100 block of Willow Avenue, my street. In particular, given the slope of the land, it was coming from the backyards on the northwestern side, where the massive white oak in the Kurtz-Greenberger yard had been cut down the year before and the Miller Tree, though still standing at the time, was effectively dead, most of its canopy gone.

Big trees act like massive water pumps, their roots sending hundreds of gallons of water per day up to the canopy leaves to be released in the air through transpiration. The leaves themselves, when storms arrive, act to slow the fall of raindrops, scattering the water with a million points of friction and deflection, till the rain drips more slowly toward the ground and the roots below. Get rid of those trees through climate change and suddenly the climate-enhanced downpours rush even faster toward the streets, hurtling toward buildings like my church, overwhelming drains.

On its own, the rainfall total was staggering on that day in 2020: nearly five inches fell in about an hour. It was a five-hundred-year storm. On lower Willow, the 7300 block, which had seen its share of lost trees, too, the water went wild, scouring front yards and flooding the basements of several houses near the end of the block, some with three feet of water. No one had seen this before either. On upper Willow, my block, the water just converged into a single stream and made a beeline for the church. It flowed from backyards, across the Tulip Avenue sidewalk, and then down the street—finding a gap in the curb at the pastor's parking space and entering the church grounds a few feet away.

Kurt, meanwhile, couldn't shake that storm from his memory.

"Taking care of a one-hundred-year-old church is like owning and taking care of a giant one-hundred-year-old house," he said the day we met. "You can imagine. So many things need fixing all the time, and you have only so much time and so much money. So when we started having water problems, it was like, 'How are we going to pay for all this?'"

Kurt spent most of his weekends doing various chores around the church. At one point as he was giving me the "Biblical flood tour," he said, "Listen, I've got to drop something off to Gary in the basement. Do you want to come?"

We found Gary Cardillo, the church building manager, dressed in a full-body hazmat suit in the basement, trying to unclog a stubborn grease buildup in the church's main drain. Kurt handed him some heavy-duty plastic bags for the debris.

"It never ends," Kurt said.

When Kurt first joined the church in 1994, there were no chronic flooding problems that he can remember. The old oil-burning furnace—vintage 1945—was a constant headache, and the slate roof was forever in need of patching. The roof rested atop the church's high granite walls, all beautifully constructed in 1922, and now presenting a mason's nightmare of cracks waiting for the next storm.

This was not an overly wealthy congregation. Like Takoma Park itself, it was racially and economically mixed, the pews full of both upper-middle-class career federal workers and newly arrived immigrant families from as far away as Africa. Also like the surrounding town, it was overwhelmingly progressive. There was a Black Lives Matter sign above the church's front door and a quote from the Reverend Dr. Martin Luther King Jr. along the northwest side of the church: "Darkness cannot drive out darkness. Only light can do that."

Around 2008, congregation members first began noticing significant changes in the hydrology around the church. Storms were getting bigger, and rain began to pool up on the Maple Avenue side of the church, where Maple and Tulip cross. The intersection would regularly become a small lake in a way no one had seen before. Eventually, the city paid to elevate the entire intersection with an additional several inches of asphalt. It was officially a "traffic calming" measure, but the change helped push the heavier rainfall volume toward the church curbs and the street drains.

Whether this was a trigger or not, the church's basement floor began conspicuously sinking in the years that followed. If the floor continued to buckle and sag, it could affect part of the church's foundation. Digging down and shoring up the floor would cost at least $100,000, Kurt said.

But that would have to wait because, also in the 2010s, whether it was just old age or more rain or both, moisture was leaking into the sanctuary and becoming a bigger and bigger problem. Finally, during the COVID pandemic, the church ponied up for a full renovation of the sanctuary and the "buttoning up of the building envelope" against water, costing $100,000.

Then, right in the middle of *that* renovation, the 2020 storm happened. Water barreled past the sloping playground and leaked into the preschool, and the city committed to the unorthodox $45,000 sidewalk berm project. But the question now was: How long would this latest solution hold? City officials say the church is safe from any storm up to and as big as the 2020 event. Anything bigger—and by definition, bigger storms *are* coming—and the Tulip Avenue engineering scheme will be overwhelmed. The small berm system would actually push future water to Maple Avenue, and the church would flood on *that* side next. And to fix *that* could cost about a million dollars, the city says. A million.

Kurt, standing once again atop the elevated church sidewalk, took a deep breath at this point in his long story. His paint-splattered flannel shirt heaved at the chest. Then he said, "It's just a matter of time, I suppose."

My US congressman, Jamie Raskin, lives on Holly Avenue in Takoma Park, two blocks from the church. He's attended town hall meetings and interfaith seders in our church assembly room—the one that's sinking.

Jamie's top reason for running for Congress in 2016 was to fight climate change. He wanted to protect his constituents from the increasingly calamitous impacts. He would frequently point out: We could cure every major disease in the world, including cancer, and still not have good health if climate change renders us food insecure. We could end all wars on the planet and not have peace if violent storms assault us everywhere.

Long before Jamie's meteoric rise in Congress—and his eventual lead role in the second impeachment trial of former president Donald Trump over the January 6 insurrection—he was stunned by the changing seasons in Takoma Park and environs. The summer camp of his childhood—Echo Hill—never shut down in summer due to weather in the 1960s and '70s. But by the 2000s, when his three kids went to the same camp, field trips to movie theaters were common, as heat-induced algae made the Chesapeake Bay toxic and unswimmable. In winter, heavy rainstorms now replaced snowfall in our neighborhood, erasing the childhood joys of snow days. It also meant the Raskin basement on Holly Avenue—like so many others—started flooding about the time his kids entered middle school, requiring costly treatment for mold.

Jamie's first job in politics was that of Maryland state senator for Takoma Park and nearby Silver Spring. For a decade, his constituents sent him to Annapolis to push for climate solutions like offshore wind power. He also helped pass a bill to ban fracking for methane gas in Maryland.

But the first time I collaborated with him to pass state legislation, it was not about climate. It was a bill to prevent the Maryland State Police from spying on peaceful environmental and social activist groups. Under former Republican governor Robert Ehrlich Jr., that had actually happened. Undercover cops infiltrated meetings run by pacifist Catholic nuns in Baltimore and groups like Amnesty International. The Chesapeake Climate Action Network was spied on, too. Jamie's legislative efforts helped end that practice.

As a constitutional lawyer, he understood two core truths: Everyone has the First Amendment right to speak out and peacefully protest. But if speech *does* move to violent action—to, say, trying to overthrow a fairly elected US government—then a crime has been committed.

When the mob of one hundred thousand anti-democracy protestors came to Washington, DC, on January 6, 2021, their goal was straightforward: attack Jamie Raskin and every other member of Congress and stop the certification of the Electoral College vote. Inconceivably, just six days earlier, Jamie had lost his beloved twenty-five-year-old son, Tommy, to suicide. If it's true that life resembles a blossoming flower—beautiful, fragile, brief—then Tommy personified that truth. He was brilliant and depressed. He was a can-do, activist-minded young man—and haunted by the injustices of this world, from animal cruelty to poverty to climate change.

That Jamie, in his grief, even made it to Congress on January 6 is amazing. He survived that day and was later asked by then House Speaker Nancy Pelosi to lead the impeachment trial of Trump in late January. Then Jamie kept going, remaining to this day a leading voice upholding 250 years of constitutional norms in this country.

Of course, the January 6 attack—at its core—was not driven by the spiking temperatures and chaos of climate change. It was driven—at its core—by the fact that lies on Twitter (now X) spread six times faster than the truth. But the larger lesson for the climate movement is still there: governments can grow ill, too. They can fall prey to disease. Like people, like trees, like the planet itself, governments have their own version of an immune system that, when attacked, can either fail or—as on January 6—narrowly survive.

For now, Jamie Raskin is able to turn more of his attention back to climate change, investigating—for example—the decades-long campaign of oil companies to deceive and confuse the public over the causes and seriousness of global warming.

But my fear is that the next big challenge to our government's health *will* come from climate change. And my small Maryland town offers a discomfiting sneak preview behind that particular curtain.

"UNFORTUNATELY, YOUR CHURCH is not the only spot in danger of flooding in this town," said Ian Chamberlain, deputy director of public works for the city of Takoma Park. "You've got Glenside, Sycamore Avenue, Elm and Poplar, and then, of course, there's Fourth Street and the whole area below Spring Park."

By "below Spring Park," he meant the neighborhood locally known as Hell's Bottom, a moniker well-deserved given recent events.

Ian and I were talking by phone, and I had just asked if the city would pay the million dollars needed to keep my church dry if a new record rainstorm came, one bigger than September 10, 2020. Such a storm would likely surpass the capacity of the new sidewalk berm on Tulip Avenue, causing rainwater to flood the church on the Maple Avenue side.

"If we ever get a rainstorm that bad—and I'm not saying we won't—then the church will have to get in line because we'll have problems all over the city," Ian said.

Those problems would include, especially, Hell's Bottom. It was one of the lowest-elevation spots in the city, where three streams converged. Prone to floods, the land values there had been historically depressed, drawing a demographically mixed neighborhood of homeowners and lower-middle-class renters. Until the 1950s, some of the houses still had outhouses, and a small pig farm hugged one streambank, according to locals. During the crack epidemic of the 1980s, it was a hideout for criminals fleeing police from the rough streets of DC. The neighborhood had earned its reputation for both hell and high water.

But it had never seen anything like the September 10 flood. Streets disappeared, sizable rocks rolled down one creek, and some flooded homes were still recovering nearly three years later.

Now the city's job was to protect Hell's Bottom from the next Big One and—more daunting—try to prevent the rest of the city from *becoming* Hell's Bottom. That job fell to forty-one-year-old Ian Chamberlain, an avid mountain biker and skateboarder who, as a kid, was always hanging out in creeks, building rock dams, fascinated by water's flow and power. After a decade working in stormwater management, he was well suited for the Takoma Park job. But honestly, it was an impossible job.

On a fine early-spring day, Ian gave me a tour of the city's half dozen most flood-prone spots, using along the way phrases like *dangerously high flow rate* and *destructive water momentum* and *nightmare scenario*.

The flooding here had become so bad it had sparked a recent feature story in *The Washington Post* where Ian's boss, public works director Daryl Braithwaite, said, "I absolutely agree that climate change is going to wreak havoc on cities like ours that are fully developed. We're really limited in terms of areas where we can provide additional storm-water management."

Those limits were why Ian had to design the elevated sidewalk at the church, whose sole purpose was to block water. And why, if that fails, the only permanent solution is to bury a large pipe from the top of Maple Avenue next to the church all the way down the street's long hill to the city's main drainage system a quarter mile away. But to bury such a pipe would require digging under and around an active gas line *and* an abandoned gas line *and* a water main *and* who knows what other pipes and structures that were down there dating back to the administration of William McKinley. Hence the million-dollar cost.

Ian drove a plain white government sedan the morning of our tour. Dressed in dirty boots, work pants, and wearing a scruffy beard, he was the little Dutch boy of Takoma Park, his fingers in many dikes. And his engineering work was impressive. Over a two-hour period, he took me to Sycamore Avenue, where he had built another elevated sidewalk berm to keep ever-rising floodwaters out of a series of homes. Then we went to Poplar Avenue, where he had somehow persuaded homeowners on a sloping lot to donate a sizable chunk of their land to the city for construction of a low, curb-like concrete dam that protected several homes below.

On we went, with Ian pointing out the many new and expensive structures the city had installed to slow down the water and combat climate change: bigger box culverts, bioretention ponds, vertical drop structures, armored stream walls, raised sections of street, and spillways.

We stopped on Elm Avenue to view brand-new water inlets along the curb. We stopped on Spring Avenue to see a high-density polyethylene pipe system. With each stop, we descended in elevation, down and down, until finally we reached the Hell's Bottom neighborhood, a collection of mostly modest pre–World War II homes and a small patch of forest.

The same five-hundred-year storm that sent water into

my church's preschool and put several feet of rainwater in basements on lower Willow Avenue—that storm caused the streets in Hell's Bottom to vanish. Phone videos taken that day show white-water rapids on Cockerille Avenue and on Westmoreland. They show manhole covers erupting into watery geysers. On nearby Second Avenue, a FedEx van got stranded in water, and the driver had to be assisted by the fire department. It was the kind of flood that can injure or kill people—but luckily, it didn't, this time.

On Cockerille Avenue, in the heart of Hell's Bottom, Ian and I got out of the car and walked over to an emplacement of large stones, known as *riprap*, under a new stormwater pipe that drained into a stream called Takoma Branch. In a storm, this pipe would spill the captured water from many of the inlets and culverts and raised sidewalks Ian had just showed me higher up in the watershed. Part of that system would slow the water down, Ian said, but would not stop the volume eventually reaching Hell's Bottom. There was now talk of building a mini dam, or weir, along a portion of nearby Second Street, creating a pond-like impoundment right here that would protect the most vulnerable homes below Second Street.

But the city had already spent $550,000 on the upstream features Ian had just shown me—all since the September 10 storm. How much would the additional weir cost? How would the city pay for it? And what if it failed? How much higher could it be built? What if the next storm is a one-thousand-year storm like the two that hit Ellicott City, Maryland, in a twenty-two-month period to our north? There was just no getting around the volume of the rain. The water had to go *somewhere*. And in a warmer world, with more and more moisture in the atmosphere every year, it kept coming and coming.

"Long term, there are only a few basic options for Takoma Park," Ian said, leading the conversation that day into one of the most gut-punching talks I've ever had about global warming. After twenty years of researching climate issues and organizing community responses, I was now confronting the endgame scenario for my own neighborhood.

Ian said the first and best option for solving the town's flooding problems was, of course, to stop climate change itself—and to do it almost immediately. There was a short pause after he said this. Then Ian grimaced and moved on to option number two.

The second option involved going ahead and building massive evacuation pipelines—from Maple Avenue, from Hell's Bottom—and leading them straight to Sligo Creek, the largest tributary in the city. But even if you could afford the millions and millions of dollars for that, a truly punishing sum, all you've done is create a giant Sligo Creek problem. You've dumped your water into a stream that will pour into nearby Prince George's County, a majority-Black county with profound water problems of its own and fewer resources. And if every community just piped its massive new water volumes into the closest main tributary, those tributaries would eventually become watery steamrollers, blowing out bridges and leveling all bankside communities in their path.

So if we can't stop climate change quickly and we can't dump our water onto the next downstream community, what do we do? I asked Ian.

"You keep the water on-site," he said. "The third option."

For the next several minutes, there in Hell's Bottom, Ian described radical infrastructure systems that would, in theory, allow the city to take the pummeling of biblical rains. Just take it. One method would be digging huge underground concrete chambers throughout the city to serve as reservoirs, storing the water belowground for gradual release into Sligo Creek after each storm. No one's even tried to calculate what that would cost for Takoma Park or to replicate it for every city in America (although DC is doing a version of this now for sewage reasons and is spending $5 billion!).

What would be less expensive, but still fantastically expensive, would be requiring every home and business in the city to have large cisterns on-site that capture rainfall and then store it temporarily in special holding tanks in existing basements for gradual release. One can barely imagine the political and financial cost of mandating cisterns and this level of intrusion into every household and commercial site in the city. Would the mayor and city council survive elections after approving such a radical plan?

Alternatively, you could pick a different but still radical approach. You could choose one or two downslope houses on every block in the city and simply knock down those houses. Tear them down. In their stead, you would turn the lots into catchment ponds for that block. It would keep the other houses dry and keep the city's water on-site. But you would have to evict families and purchase homes and reduce the housing stock in a city already sorely lacking enough affordable units.

Again, could local politicians survive that? Probably not. But could they and the city bureaucracy survive doing *nothing* when the next five-hundred-year and one-thousand-year storms come? And come? Could the city's functional immune system handle any of these options without collapse?

Right then, thinking through the options with Ian, I

reached a state of palpable panic. This was the end of the road for pondering *What if?* for my community's future. In practice, we can replant many of the trees that have perished—and hope the new ones survive. And for illnesses like Lyme disease, we can wait for possible vaccines and other treatments even as the world warms. But the rain? There's no real solution here, no way to hope it away or outwait it. It is, as those climate scientists wrote in their February letter, our version of "climate change impacts beyond humans' ability to adapt." Every place on the planet has its own version of the endgame, of circumstances that can't plausibly be overcome. California has fires. Miami has sea-level rise. Central China has drought. Takoma Park has storms like that of September 10, 2020.

For a long time after that conversation with Ian Chamberlain, there in Hell's Bottom, I was depressed again. We had passed many rooftop solar panels on that springtime tour, at least one on every block. And electric cars were charging in many driveways. But like the eighty-one solar panels on my church's roof, it wasn't a solution for right now or for the near future.

I thanked Ian for his time and returned to my everyday life, thinking, *When will my own house flood? When will the preschool close at the church? Is the next big storm coming next week? Next month?*

When I was younger, before I became a climate activist, I had recurring nightmares about not preparing for college exams or being trapped in crashing airplanes. But now I mostly have nightmares about water.

My most common dream, one that continued after meeting Ian, is this one:

A giant rainstorm comes during the day, but it feels like

nighttime because it's so dark outside. The rain pours and pours for hours. The sound is deafening, and I can't see my street, because it's completely underwater. Like a river, the water moves past, barreling toward Tulip Avenue, disappearing toward the church.

Beth and I can't leave the house. We open the door down to the basement, and water gushes onto our feet. It's reaching the first floor. We call the fire department. We call the police department. They don't answer. They're all out helping other people and dealing with their own flooded buildings. Government services are gone.

We go to the attic and wait for a long time until the rain finally stops. Then we go outside, and we see the roads have all been washed away and there are no bridges across the streams and creeks of Takoma Park. And we stand there. And we don't know what to do.

THE DAY JAMIE RASKIN arrived at my front door on Willow Avenue, it definitely wasn't a dream. It was March 25, 2023. He had lost seven pounds, was wearing a bandanna covering his balding head, and he had no eyelashes. The chemotherapy had been hard on him.

I had contacted Jamie weeks earlier, asking how he was doing and whether his treatment for diffuse small B-cell lymphoma cancer might still allow him to take a short walk around the neighborhood. I knew he missed, as I did, our occasional strolls to discuss legislative matters on Capitol Hill and to catch up on climate change issues. I also badly needed a dose of his famous optimism. "Let's shoot for around March 25," he said. "My T-cell count and hemoglobin numbers will be at their best levels in the treatment cycle. I think I can walk then."

Sure enough, on a gray, chilly Saturday, he appeared at my door, the man who had done more than anyone else to take on Donald Trump and beat back American fascism and who had tragically lost a son along the way and who was now four out of six treatments into chemotherapy for cancer at age sixty—that man was at my door. And he was carrying a cake. "I thought you and Beth might like this," he said, handing me the chocolate swirl coffee cake.

I don't remember what I said at that stunned moment. I wanted to bear-hug his frail frame. I shook his gloved hand instead, given his immune-compromised state. I put the cake away, and we sat on the front porch as Jamie rested from the three-block walk to get here. "The doctors say the cancer is gone," he said. "It's not in my body anymore. But I still have to finish the last two rounds of chemo." He had been diagnosed in December, but still kept up most of his work on Capitol Hill somehow and his many appearances on cable TV via Zoom.

We chatted a bit until he insisted he felt fine to keep walking. We headed down Willow Avenue toward the DC line, taking our time, making our way toward the two prized Yoshino cherry trees at the end of the block. I wanted to show off that year's magnificent bloom to him.

DC park rangers and embarrassed TV weather forecasters will tell you that predicting the precise date of peak bloom for these fickle cherries is nearly impossible. So I took it as a good omen that the date Jamie had given me nearly a month ago for our walk now corresponded to the precise first day of peak bloom for these Willow Avenue cherries. The Yoshinos around the Tidal Basin had peaked two days earlier.

We paused for a moment under the first tree, its branches arching over the sidewalk with a million white-and-pink petals. Jamie pulled down a branch and breathed in its fragrance, a ritual gesture of spring. But in his condition, it seemed like so much more. I saw a man on chemotherapy breathing in a thousand years of healing Buddhist guidance: Life is beautiful. Life is fragile. Life is brief. Be well today. Be happy.

And he *was* happy that day. "It's so good to be out. What a wonderful spring day."

Of course, he was putting a positive spin on the day itself. There was no sun. It was gray. It was fifty-nine degrees at the tail end of a chilly late March—after a near-record warm January and February. Normal weather: we just don't have it anymore. But Jamie, for all the tribulations he had endured the past two years, enough to fill ten lifetimes, was always stubbornly upbeat, a self-described "serial optimist." One of his congressional staffers called him "Dr. No Problem." Against all odds, there was always a ray of sunshine glowing bright in Jamie Raskin's view of the world.

"It's so good to see you, dear Mike," he kept saying.

Still, there was hard reality. On his way to my house, he had passed the elevated sidewalk at my church. And he had walked near the sad remains of the Millers' dead red oak. And on my porch, he had told me how on his street, too, Holly Avenue, they had lost "many, many trees" in recent years and his own roof was being repaired for recent storm damage from falling branches. In turn, I told him about my recent tour of Takoma Park flood spots with a city official and how it had left me feeling, frankly, nearly hopeless.

So I posed the question that had motivated me most to reach out to Jamie, this congressman who had run for office to fight climate change above all.

"How do you do it?" I asked. "How do you stay optimistic about climate change with everything happening right here in our own neighborhood? How can *anyone* stay optimistic?"

He paused for a moment there on the sidewalk. He stared down at his feet. On the ground, there were almost no cherry tree petals despite literally millions above us. I glanced up. The branches were drooping with their cargo, the flowers soothingly pale pink against the pale gray sky, a pastel sublimity.

"What keeps me optimistic?" he said. "Maybe this sounds strange, but I think it's the trees themselves. We've lost so many, but when one falls, so much sunlight comes in that we see what we're missing and then new trees come and we start over again."

He continued, "Like the whole era with Trump. We've had setbacks, but they have immediately invited new scrutiny, new light, so we can see how important democracy is, perhaps more than ever."

Jamie kept talking as we slowly pulled away from the gorgeous cherry trees. We made a U-turn and began walking under the big willow oaks that line the sidewalk opposite my house. They were survivors of all the recent extreme weather, these trees. Nine of them in a row along the block, still big and seemingly healthy.

He continued, "The best metaphor, really, is the roots," he said. "I gave a speech once to a group of people in Rock Creek Park, in the forest, and I said it was what you can't see, underground, that keeps this forest and our nation alive. The many roots, interwoven, keeping us connected and strong and nourished. We've lost a lot of trees, for sure, but look at the ones still here in Takoma Park, the roots still working even when we can't see them. Most Americans want to save our planet. They want a free society and a sustainable economy. Those are our roots. We've got a lot of challenges ahead and we don't have all the answers, but those are our roots. And that's what's going to save us whatever storms come our way."

We turned onto Tulip, making our way back toward Jamie's house on Holly. The street was lined with much younger Yoshinos, spaced twenty feet apart, the canopies smaller but still ostentatious and peaking.

Without prompting, Jamie brought up his son, Tommy, who had died by suicide in 2020.

"You know, Tommy used to ask the question, 'Given the climate emergency, given all the political polarization and hate, given how many people are already on this planet, is it *ethical* to even have children?'

"It's a valid question. But we used to tell him, 'Tommy, you came into this world, and you made your parents better people and made your community better and the world better with your activism and presence. We're sure glad we had you.'"

He had stopped walking at this moment to punctuate that last point. We were almost to his house. It was the only moment during our walk that he seemed tired to the point of exhaustion. He turned to me, scarf on his head, cable-knit sweater buttoned tight around him, his face slightly puffy from the chemo. "So that's why, despite everything, I recommend to all the young people in my family that they have children. They should bring babies into the world. So yes, I think that makes me optimistic. I have hope."

In all of Takoma Park, the most beautiful stand of cherry trees is actually on Jamie's street, Holly Avenue. The Yoshinos are much older, taller, and decidedly pinker there on Holly's upper stretch, near Eastern Avenue. Jamie and I stopped to take a selfie with the sea of blossoms behind us.

Then I walked him to his front porch, past the old-fashioned picket fence of his 1890s home. He waved goodbye at his doorway, smiling below the roof of that old house, whose shingles had just been repaired from all the recent storms and fallen branches.

5

An Act of God

March–April

I was at Ning's house one evening in mid-March when the issue of Mount Pinatubo came up. I was touring his basement laboratory, viewing the microscopes he used to check wood samples buried under different soil types. He had just showed me a chunk of the 3,700-year-old cedar tree his team accidentally uncovered near Montreal.

"Mount Pinatubo was an act of god," Ning said. "And there's a big difference between an act of god—and human beings *acting* like god themselves."

He was referring to the massive volcanic eruption in the Philippines in June 1991—Mount Pinatubo. It spewed ash and dust and twenty million tons of sulfur dioxide into the atmosphere. Since airborne sulfur has the quality of reflecting sunlight, the eruption resulted in an astonishing meteorological fact: the average temperature of planet Earth dropped by one degree Fahrenheit for more than a year after the 1991 explosion. Some cooling continued for months after that, tapering off. The scattered sulfur from the eruption was faintly dimming the sky, blocking out part of the sun. Ever since, many climate scientists have wondered, *What if . . . ?* What if, in a controlled way and with prudent safeguards, we could become a "human volcano"? What if we could disperse enough sulfur high enough in the stratosphere to reflect just 1 to 2 percent of the sun's incoming rays? That would temporarily cool the planet long enough for us to get off fossil fuels for good.

I brought up the issue of Mount Pinatubo that night at Ning's house because, again, I was growing increasingly despondent. Having dug deeper into the relentless water problems of Takoma Park and discovered my town was not likely to adapt to that challenge, I remained preoccupied with the February letter from James Hansen and other climate scientists calling for geoengineering research on a significant scale. Unspoken in that letter, but underpinning its argument, was the eruption of Mount Pinatubo. The dynamics and consequences of that explosion had been thoroughly documented by science. And if there was a hint of an escape hatch in this overheating world—à la Jamie Raskin's "we don't have all the answers . . . but I have hope" remarks—maybe it was Mount Pinatubo.

In 1991, global satellites and balloon spectrometers and terrestrial weather stations recorded the remarkable details of the Pinatubo eruption. The cooling lasted for more than two years in total as the sulfur aerosols gradually dissipated in the atmosphere. In the meantime, extremely mild summers were recorded from Australia to Scandinavia. Jerusalem recorded its biggest snowstorm in forty years in 1992. And the polar ice caps actually *grew* in 1992 and 1993, expanding in mass balance by over four hundred kilograms per square meter at the peak.

The US East Coast, meanwhile, was not exempt. The winter of 1992–93 brought the so-called Storm of the Century, a nor'easter dumping record snow across the Eastern Seaboard. And the coldest stretch of winter weather I can ever remember in thirty years in Takoma Park started in late 1993. On Willow Avenue, during one stretch, the low temperature dropped to near or below zero (unheard of now) for several days in a row that winter. All the neighborhood fig trees died, and frozen water pipes burst in my basement for the first and only time. Heating oil ran out across the region. My back porch was so frozen that taking one step onto the wooden boards sent a loud cracking sound rippling across the floor like ice on a frozen pond. It hasn't been that cold here since. While direct attribution to the volcano is impossible, the weather on Willow Avenue for a while matched the global pattern.

In short, in an unanticipated way, Mount Pinatubo ran a dramatic climate test for the planet. It erased, for an extended period, a century's worth of man-made global warming that had accumulated prior to the explosion. But by 1994, the warming returned to pretty much where it had left off in 1991.

Yet all these years later, sitting in his home laboratory, climate scientist Ning Zeng said the concept of solar radiation modification made him nervous. There were too many things that could go wrong with humans cooling the planet in a major way, he said. Too much we don't know.

"So isn't that a good reason to study it as fast as we can?" I asked Ning. "To learn more?"

"Yes, I think so," he said. "I support that. But you should go talk to my former student. His name is Tianle Yuan. He signed the letter with James Hansen calling for geoengineering studies. He works at Building 33."

"Building 33?" I said. "You mean at NASA?"

"Yes, that building. I did my postdoctorate work there. You should talk to the people at NASA."

Then he paused and said, "I'm going to take you to Building 33."

THE IDEA THAT volcanic eruptions can cool the earth was not a new concept. Ben Franklin, living in Paris in 1783, theorized that an eruption in Iceland had caused the haze over Europe that year and triggered the exceptionally cold European winter of 1783–84. Later, the eruption of Mount Tambora in Indonesia in 1815 was followed by "the year without a summer" in much of the Northern Hemisphere, where August frost killed crops across New England.

In 1965, scientists working for President Lyndon B. Johnson were likely well aware of the volcano theory. But no one understood the precise atmospheric chemistry of the phenomenon, so the global warming report presented to Johnson proposed only attempts to modify cloud behavior and place reflective floating particles in the ocean.

Then, in 1974, respected Soviet scientist Mikhail Budyko theorized that it was not the ash or dust but the sulfur spewed from volcanoes that was the main source of the cooling. Again, sulfur aerosols (tiny water droplets in the air suffused with sulfuric acid) have the property of scattering

AN ACT OF GOD

and reflecting some of the sun's incoming rays back into space. To counteract planetary warming, already detectable by 1974, Budyko proposed using human-launched balloons or airplanes to release sulfur dioxide into the stratosphere "much like that which arises from volcanic eruptions."

For years, it was just an untested theory. Then, on June 15, 1991, Mount Pinatubo happened. It exploded with the force of thirteen thousand Hiroshima bombs. Villagers living on the mountain's slopes didn't even know it was a volcano. It had been dormant for five hundred years.

Plumes of ash and rivers of molten rock eventually killed 722 people and left 200,000 homeless. Meanwhile, a dense cloud of sulfur dioxide—250 miles wide—entered the stratosphere, 13 miles up. From there, it began to spread, driven by powerful winds, around the world. Within weeks of the eruption, the entire planet had begun to cool.

And NASA scientists saw it all.

NING AND I flashed our temporary badges to a pair of armed guards at the gate. "Go directly to Building 33," the guards said, waving us past a tall perimeter fence topped with barbed wire. Suddenly, we were there, inside the grounds of NASA's Goddard Space Flight Center.

My Hyundai electric car showed we had traveled just eight miles from Takoma Park, yet I had never been to this 1,300-acre facility. It sat just outside the DC Beltway, in Greenbelt, Maryland. No one I knew had been here either, until I met Ning. The place kept an amazingly low profile amid high security.

Now, in a campus setting of green lawns and rambling forests, we were passing secretive-looking buildings, large and windowless, where America's pioneering space rockets were developed to send early US satellites into orbit. Established in 1961, in the wake of the Soviet Sputnik launch of 1956 (the world's first orbital satellite), the campus today has about ten thousand employees with a hand in fifty active space missions. Those include the James Webb orbiting telescope, two Martian probes, and the Parker Solar Probe, which makes periodic close passes by the sun.

So large is this campus that Ning, who's been here many times since his student days, quickly got us lost. After a few bad turns, we came to an angular brick building of drab '70s-style architecture. It was four stories tall, with large hidden corridors in the back and a wide parking lot in front, tucked in a far corner of the Goddard property. Above the front door, in blunt metal script, was the number 33. No fancy names here among the scientists. Visible on the roof were various antennas and mechanical arms and other strange-looking machines, some of them moving and seeming to scan the sky. This was Building 33.

Exiting the car, Ning first pointed to the thick forest of hardwoods and pines surrounding the building. These woods had inspired his early tree-burial ideas, he reminded me. "Let's check it out," he said.

We walked one hundred feet to the edge of the parking lot, and then disappeared through a curtain of trees, entering the woods that had changed Ning's life. We walked along a loose trail, leaves stirring below our feet, and then paused, gazing silently at the mature oaks and beeches and white pines all around. In the dappled light of this placid spot, we began to feel the stress of the morning Beltway traffic and other cares fall away. It was a spontaneous moment of *shinrin-yoku*, of forest bathing, translated from Japanese simply as "soaking up the forest atmosphere." The results were immediate amid these straight trunks and arching boughs. Had we been hooked up to monitoring equipment, we would have watched our cortisol levels—our stress hormones—steadily fall at the same time our heart rates dropped. We were internalizing the stillness of the trees. The feeling was wonderful.

It was the last week of March and winter had officially ended, and these hardwoods and pines would soon be cranking out legions of flowers and clouds of pollen themselves, dusting the region with their annual yellow-green film that clings annoyingly to every surface by late April. Leaving the woods moments later, thinking about the coming change in seasons, I could feel my stress levels climb right back up, nearly as fast as they had fallen. The reason: full-on spring was almost here, and we had never really *had* a winter. Temperatures had dipped below freezing several times, but there had been no hard-frozen earth crunching below our feet. And no snow had accumulated beyond trace amounts across this perpetually thawed region. Indeed, we were now in the middle of what would become a two-year "snow drought"— the longest ever in many parts of the DC area.

All of which made me eager to visit Building 33, chasing new conversations in pursuit of a possible plan B for this increasingly unrecognizable region and planet. As Ning and I crossed the parking lot, heading toward the building entrance, he told me again that this sprawling facility housed one of the greatest concentrations of earth scientists in the world. Men and women who study oceans, glaciers, hurricanes, forests, volcanoes, the atmosphere, precipitation, polar ice, droughts, heat waves, and on and on—they worked here with legendary multidisciplinary coordination. After winning the space race of the 1960s and deploying pioneering satellites for mostly communication purposes, NASA by the 1970s began peering down at an obviously fragile and troubled home planet in need of exploration itself. So Building 33 was eventually born, a kind of mini Pentagon of earth scientists.

The moment you enter the building, here's what you see: satellites. They're everywhere. Not real ones but replicas behind glass or photos of satellites on the walls. Each one includes text explaining how satellite X is feeding data Y to research program Z—using space-based cameras, lasers, and other measuring devices.

Which is how Ning and I got immediately punched in the gut the day we entered Building 33. Right inside the entrance, at the first display, was a time-lapse series of satellite photos showing the meltdown of the Zachariae Isstrøm ice sheet in northeast Greenland from 1999 to 2022. Playing in a loop, over and over again, the photos showed how a buffer of floating coastal ice had shattered like glass and begun melting rapidly in 2002. This unlocked huge masses of land-based Greenland ice, behind the buffer, that began sliding toward the open ocean throughout the 2010s. This one phenomenon, as narrated by the display, could raise the planet's oceans one and a half feet worldwide this century.

Ning and I stood there for a moment, both silent, taking in the horrifying images.

"This," Ning finally said, "is why I had to do more than just teach climate science."

He pointed to the photos of disintegrating Greenland ice again. "The alarm bells are screaming at us now. I had to *do* something. Bury trees. Store carbon. Do something."

In that context, Ning had invited me to attend a lecture on federal forest management issues at Building 33. But the lecture was hours away and Ning had other meetings scheduled, so I had time to kill. My goal was to find his former student Dr. Tianle Yuan, who was an expert on how clouds and aerosols reflect sunlight away from the planet. I had failed to connect with him by email and was going to try in person.

Like the Goddard campus itself, it's easy to get lost in Building 33. I informally wandered through multiple floors on eight different building wings, armed with my Goddard badge clipped to my fleece vest, passing a steady stream of hall traffic. Different wings housed different concentrations of NASA earth science operations: Atmospheric Chemistry and Dynamics, E3 and E4. Global Precipitation Measurement, G4. Ocean Ecology Laboratory, H6. Global Modeling and Assimilation Office, C2. Most programs displayed a photo of a satellite on their hallway walls, explaining how key data was relayed from space to fulfill their earth mission.

Talking to people in the halls, as I did, curious about their programs and asking for directions, I encountered an amiable string of researchers and technicians speaking a mysterious language full of government acronyms—SMAP, PACE, SWOT, OCO-2, HBG—as well as frequent references to every imaginable *-sphere*—lithosphere, hydrosphere, biosphere, stratosphere, mesosphere. Forget outer space, it's a whole different world right here in this building.

When I finally found Tianle's office—305A, Joint Center for Earth Systems Technology—he was not there. One of his colleagues gave me his phone number and asked what my interest was. I told him I wanted to learn more about how sulfur dioxide released by volcanoes and by human activity can cool the planet.

"Oh, you want to go visit AERONET," he said. "They measure aerosols from the troposphere to the stratosphere."

I didn't quite understand, but I took directions and got

lost again and then ascended a long flight of stairs to an open door that led to what looked like a machine shop. Metal racks held an assortment of parts, some with tags on them, atop a concrete floor.

"Aerosol Robotic Network. That's what AERONET stands for," said a man, who looked to be in his twenties, seated closest to the door. He turned away from a large computer monitor to peer at me. "You didn't see all the robots on the roof when you pulled up?"

I told Arsenio Menendez, the AERONET systems engineer, that yes, I had seen strange objects on the roof of Building 33, but I didn't know at first they were robots.

He suddenly tapped his keyboard to pull up a map of the world's continents on his monitor. Every continent, including Antarctica, had small red dots on it, some in bunches, some spread out, but dots everywhere, worldwide.

"Each one of those spots is a set of robots that my hands have touched," said Arsenio. "We calibrate them and repair them here. They use photometers to scan the sun and atmosphere and measure smoke, dust, sea salt, sulfur particles, and other pollution from here to the top of the stratosphere. See that dot right there?" He pointed to the red smudge just outside of DC in North America. "That's this building. Those robots are right through those doors over there—on the roof."

From Arsenio and others in the department I interviewed later, I learned that, for all their wonders, satellites can't do it all. Their measurements from space need corroboration wherever possible from instruments on the ground. If you read the fine print of most scientific studies using satellite measurements, you'll see a reference to the AERONET data offering ground-based confirmation.

And what both of these systems have been saying for

years, I learned, in study after study, from space and from the ground, using satellites and photometer robots, is that human beings have actually been *cooling* the planet—for decades. We've *already* been filling the atmosphere with aerosols and reflecting sunlight away from the earth—accidentally. It's been happening at the same time we've been warming the world at an even faster rate. It's a complex concept, hard to wrap your brain around at first—and at that moment at NASA, I had a decidedly simpler thought on my mind. I wanted Arsenio to take me onto the roof of Building 33 and show me the sky-probing robots.

He wouldn't do it.

"You have to have special permission for that," he said. "You have to have an escort."

A FEW WEEKS later, I returned to the Goddard campus by myself. I got my badge at the front entrance, got promptly lost on campus, then reached a small conference room in Wing A of Building 33. The room was replete with filing cabinets, a whiteboard, and the requisite satellite wall photo with lots of wonky data. Outside, a new heat wave had returned to the region in mid-April, with temperatures reaching the high seventies many days, supercharging the arrival of spring and the emerging flowers of oaks and other trees in the forests around Building 33.

I had come to finally meet Dr. Tianle Yuan. On a brief call the day before, he had confirmed that, as a graduate student at the University of Maryland in the early 2000s, he had taken a global warming course called Advanced Climate Modeling. It was taught by the dynamic young professor Ning Zeng, a rising star on the campus.

Now, twenty years later, Tianle was a senior researcher at NASA. When he entered the conference room that day, he greeted me in a soft Cantonese accent, having grown up in Jiangsu Province in China, just north of Shanghai. He had traveled to the US as a young man for an advanced degree.

In Ning's modeling class, Tianle said, he learned that, yes, carbon dioxide is released whenever fossil fuels are combusted—and a warming planet is the result. But those same fossil fuels—oil, coal, and methane gas—also release sulfur dioxide when burned. The sulfur concentration varies from fuel to fuel—with coal having the most. But the collective consequence of combusting carbon fuels worldwide every day is the creation of an ongoing aerosol blanket around the globe. That blanket reflects sunlight and reduces *by a full one-third* the extent of the warming we would otherwise experience on Earth, according to a recent analysis. The planet, super-warm already, would be significantly warmer without this mask, in other words.

"So doesn't that make us—human beings—a kind of Mount Pinatubo already?" I asked Tianle there in the NASA conference room.

"Are we," I continued, "a 'human volcano' on a smaller scale? Pretty much all the time?"

"I suppose so, yes, we're kind of like a volcano," said Tianle. "The question now is: Can we figure out ways to keep reflecting sunlight away from the earth in a safe way even as we reduce fossil fuel use? We need to try to figure this out."

Tianle looked much younger than his age, forty-two. He had long black hair pulled back in a bun and a swimmer's body from hitting the pool three times per week. By all indications, he was also something of a genius, having just finished a project using artificial intelligence to better predict the intensity of hurricanes. Now he was turning his attention to the issue of geoengineering.

His main focus was something called *marine cloud brightening*—that is, studying ways to artificially increase the whiteness of stratocumulus clouds and trade cumulus clouds over the world's oceans. Done on a large scale, this could increase the planet's reflectivity. Some proponents envision solar-powered drone ships traveling the oceans, enhancing clouds with a saltwater mist sprayed constantly into the sky. But the science around cloud brightening is exceedingly complex, Tianle confessed. There had been some progress in research since the 1990s but no giant breakthroughs since a version of cloud modification was first proposed in the 1965 Lyndon Johnson global warming report. Recent pioneering attempts to brighten clouds over the Great Barrier Reef in Australia have been inconclusive, for example.

What *was* more likely to work—and was better understood thanks to Mount Pinatubo's natural experiment—was seeding the stratosphere with sulfur particles.

"Of course it will work," Tianle said. "Of course it will cool the planet if we do this. But what could be the unintended consequences? How will it affect the ozone layer, for example? That's why we need to study it and do field research now."

He was the only scientist here at Goddard to sign the Hansen letter calling for major geoengineering research. There was definitely a rising interest in the issue on the campus, Tianle said. But many scientists prefer to avoid any appearance of involvement in policy or political issues. Others fear even the study of climate-altering techniques will lead to overconfidence and potentially reckless policies.

"But for me, I just don't think it's controversial to want to study this anymore," Tianle said. "The larger public, I think, has no idea how much warming is coming, especially when we take off the mask."

The mask, again, is the dome of sulfur aerosols created in the lower atmosphere when we burn fossil fuels. That sulfur, once emitted, washes out of the sky relatively quickly. But the CO_2 from carbon combustion stays in the atmosphere for a century or more. So the warming—turning neighborhoods like mine upside down with dying trees and flooded buildings—will continue even as we transition to clean energy. That is, again, unless we figure out a way to suck CO_2 out of the air—or we reflect sunlight.

Tianle had a pair of white-rimmed sunglasses propped on his head that morning—for a reason.

"You want to see the robots?" he asked, pointing upward. "You ready to go on the roof?"

"Very much so," I replied, and into the Building 33 labyrinth we went, working our way toward the AERONET shop. Along the way, I asked more geoengineering questions of Tianle. To disperse sulfur dioxide way up in the stratosphere, where it will have a longer lifespan before dissipating, most experts agree humans would have to construct hundreds of specialized planes with large fuselages to store the sulfur—and with 150-foot-long wingspans to disperse it. The planes will have to make a steadily growing number of flights each year—for decades—putting as much as one million tons of sulfur dioxide in the sky annually at the program's peak. How did Tianle respond to critics, some of them noted scientists in this very building, who say this is scary as hell? It's the ultimate act of humans playing "techno god" with the planet.

But before Tianle could answer, we reached the AER-

ONET shop at the top of Wing G. We paused at a set of double steel doors leading to the roof while Tianle pressed his security badge to a wall scanner. The doors unlocked—and we stepped onto the roof.

Immediately, all around us, were countless moving objects. The first objects I noticed—straight ahead—were actually trees, not machines. The wind had picked up, and the awakening trees four stories below were shaking gently, thousands of them, in a soft morning breeze. Dressed in early April hues of green and gold, the trees leaned this way and that. In the distance, amid the trees, was the rest of the NASA campus—the testing facilities, the office buildings, the water tower. Farther out still, over the horizon, was Takoma Park to the northwest and Washington, DC, due west. The sky was light blue and the sun strong and bright.

The second thing I noticed in motion were the robots, dozens of them on the roof, to our left and right. Each was about three feet tall with a stationary, stalk-like body attached to the roof. The "head" was a long, double-cylinder sun photometer, a kind of camera, that swiveled up and down and all around, pausing to look at the sky from multiple angles, governed by a preset algorithm. YouTube offers details on how these things work, but the basics are these: Every few seconds, the robot turns directly to the sun, records its optical radiation intensity, then pivots to a set of other points in the sky, pausing to measure the quality of sunlight there, then returning back to the sun, starting the process over again. From the ground to the stratosphere, if there is anything affecting the quality of the sunlight—forest fire smoke, volcanic debris, sulfur from coal-burning power plants—it gets measured. At hundreds of locations on seven continents, these robots help produce a composite picture of the aerosol blanket created by both natural events and, significantly, human-caused pollution.

It was easy to feel a little dizzy up there on the Building 33 roof that morning. Spring was unfolding all around us, with a pair of white-breasted nuthatches sneaking into a nearby tree, straw in their beaks for a hidden nest. I'd seen them from the ground, too, as I entered the building. Meanwhile, the many oak trees that dominate these Goddard forests and the regional landscape beyond were prepping for reproduction, too, their long and stringy catkin flowers now emerging, producing pollen that was spread not by insects but by the spring breezes like the one kicking up now. The unusual heat that month, returning with galling excess for weeks, would trigger a pollen season like few others, with April 13 posting 3,139 pollen grains of pollen per cubic meter of air, the highest count of any day here in thirteen years.

But now, strangely, up there on the roof, the ascending spring sun, which should have warmed this scene, was blunted by an odd feeling of coldness coming from the AERONET robots spread about. Their metal bodies, draped with hanging black cables, kept moving silently in our presence, gathering information soon to be checked against the satellites, themselves far, far overhead at that same moment. And one day, somewhere below those satellites overhead and above these photometer robots on the ground, could come a fleet of stratospheric aircraft emitting sulfur every day for the rest of our lives and our children's lives, those releases checked constantly by the busy robots and the space orbiters.

"Does any of that just make you scared, like some of your colleagues?" I asked Tianle, returning to my geoengineering questions. "Are we taking on more than humans can handle, playing god?" Critics, I reminded him, pointed to humans' poor cooperation and management of the environment at the planet's surface. Why would we do better in the stratosphere?

The morning sun was growing brighter, and Tianle put on his sunglasses just then. He was wearing a thin gray hoodie, the same one he wore to his ten-year-old son's soccer games on Saturday mornings. The father and son had just returned from a spring break trip to New York City, visiting museums.

"Nobody wants to put particles in the atmosphere," Tianle said. "When I agreed to add my voice to other scientists speaking out on this issue, we made sure to say right at the top that we don't want to do this. We don't want to engineer the climate. Nobody does. And we're not saying we're committed to it now. We just want to seriously study it. Create the models. Put money into research. Figure out some safe ways to test it as much as we can while creating transparency and good global decision-making—all before any decision is made.

"But here's what I think: if we are powerful enough to change the whole planet with global warming, we are powerful enough to stabilize the planet, too. There have been lots of unintended negative consequences from all kinds of technologies in the past. But honestly, in the past, we didn't really have the mindset of 'What can go wrong?' We knew less then. Now we know more. Now we can study things and try to reduce the chances of bad consequences. We didn't do that when we started using fossil fuels. We didn't even know to think that way."

There was a flight of stairs on the roof leading up to a platform holding still more photometers. We walked up the stairs, stopping at the highest point while I took photos and Tianle looked out across the Goddard campus and all the swaying trees.

"There is a Chinese saying," he said. "It goes, 'If you know yourself, you're already wise.' Do we humans know ourselves completely? No. But we know ourselves—I think—better than in the past. We're a little bit wiser because we know we're fallible. Honestly, this makes me more confident we can solve these problems now."

Then Tianle announced it was time to go. He had a paper to edit on the sulfur emissions from cargo ships out at sea. I took a few more photos, and we descended the stairs and passed through the double doors back into Building 33.

For all the talk of god and godlike powers in conversations about geoengineering, the actual process of seeding the atmosphere with sulfur aerosols is pretty straightforward—and relatively cheap. And that's part of the challenge. China, Brazil, the United States—they could all geoengineer the whole planet pretty quickly if they wanted to, on their own, with a small fleet of modified aircraft and an easy-to-get supply of sulfur aerosols. Elon Musk could do it. Shockingly, there's no international treaty explicitly forbidding any country or individual from doing this.

Some scientists have proposed key steps for moving forward. First, adopt an immediate global ban on large-scale deployment of geoengineering techniques by *any*body—while simultaneously launching the proposed multiyear, well-funded study into the best and lowest-risk ways to potentially proceed. Then, after a decade of presumed intense diplomacy, finalize an international decision-making process and then decide—yes or no—on proposed cooling techniques.

The stratosphere is not the only focus. We may one day see large fields of ground-based mirrors across the planet reflecting sunlight. We may also see much of the man-made world—roofs, roads, airport tarmacs—covered with white paint to replace some of the reflectivity lost to polar ice melt. And Tianle and others still hold out hope of techniques to brighten marine clouds. But stratospheric intervention is the leading candidate for creating a more uniform reflectivity at a combined cost estimate of between $250 billion and $2.5 trillion through the year 2100. That cost is not peanuts, but even at the high end, the yearly average expense is about one-twentieth of the annual budget of the Pentagon.

A special geoengineering research program at Harvard University, funded in part by Bill Gates, has put a lot of brainpower into the logistics of solar radiation modification. Scientists there like Dr. David Keith have estimated that as few as ten airplanes—again with special fuselage storage capacity and large wings for releasing plumes of aerosols—would be needed in the first decade or so. Those planes would release about 250,000 tons of sulfur per year in the early phase. The amount, and the number of planes needed to disperse it, would grow decade by decade to counteract the still-mounting momentum of CO_2 in the atmosphere during the waning days of the fossil fuel era. But late in the century, after fossil fuel use has disappeared and methods like burying trees and capturing carbon directly from the air have matured, those flights would begin to ramp down, peaking at around one hundred planes delivering about 12.5 million tons of sulfur per year. (Mount Pinatubo, by comparison, emitted 20 million tons in 1991.)

Customized planes circling the planet daily to keep the oceans from rising? The idea will take time to accept, for sure. But again, we've already been doing a version of this since the dawn of the industrial age, just accidentally and sloppily and haphazardly through the heat-masking effects of sulfur dioxide mixed into our myriad fossil fuel uses in the lower atmosphere.

Still, it's no wonder that people—religious or not—invoke the concept of god so regularly when talking about geoengineering. The ethical questions—much more than logistical—are daunting: Who gets final say over whether and how to proceed? How exactly will an international decision-making body work? Will the poorest and most vulnerable nations have full input? And what about the so-called moral hazard of geoengineering (i.e., the risk of excusing continued fossil fuel use)? Why get off dirty energy if we can eliminate the warming by-product with sulfur? And perhaps most troubling, what about the risk of termination shock? What if the stratospheric flights suddenly stopped at some future date due to political strife or war or natural disasters? The planet could jarringly snap back to a high temperature state in just a few years, potentially making adaptation extremely difficult if not impossible.

Then there's the issue of the ozone layer. Some of the injected sulfates will eventually affect the ozone molecules (three oxygen atoms bound together as O_3) in the atmosphere's ozone belt. The sulfates would cause some of those molecules to break down and reduce the radiation-blocking benefits so vital to the planet. Yet Paul Crutzen, the Dutch scientist whose research in the 1970s helped establish that human activities were damaging the ozone layer (for which he later earned a Nobel Prize), was a strong proponent of research and testing of stratospheric geoengineering before his death in 2021. Crutzen's own calculations showed the impact of sulfur injection would likely be minimal and temporary.

David Keith has proposed many ideas to reduce other

risks of stratospheric intervention, born of the hard-won understanding that all human technology comes with unwanted consequences, some of them huge. First among Keith's recommendations: ramp up the stratospheric flights very gradually and carefully. If there are early-emerging surprises and unintended consequences, you can shut the program down without significant termination shock. If the impact on the ozone layer is bigger than the computer modeling and field testing predicted you can shut it down. If, as some fear, geoengineering could cause the South Asian monsoon—vital to agriculture for 1.5 billion people—to grow erratic or shut down, then shut down the geoengineering.

Further reducing the risk—physical and moral—is Keith's proposal that the world's nations aim to reduce only *half* the planet's warming through stratospheric aerosol injection. Today, our planet is on course to warm at least 2.9 degrees Celsius (5.2 degrees Fahrenheit) by 2100, a disaster for nearly all living things. A cooling program aimed at cutting that figure in half would likely avoid calamitous climate impacts while using only half the stratospheric sulfur *and* maintaining a strong incentive to further repair the planet (and its acidifying oceans) by getting off fossil fuels entirely. (Interestingly, because sulfur emissions from fossil fuel use have been dropping worldwide in recent decades thanks to smokestack scrubbers and other means, the planet would not likely see a net increase—above today's levels—of sulfur deposits in the world's soil and water from deployment of this geoengineering technique, according to a 2020 study.)

But for now, the goal is simply to invest—in a big way—in more research. According to Tianle and his colleagues, we need over the next decade a dramatic increase in computer model simulations, analytical studies, small-scale field experiments, and accompanying observations.

Surprisingly, and with little fanfare, the US Congress in 2020 leaned slightly in this direction by approving an annual budget of $10 million for the National Oceanic and Atmospheric Administration (NOAA). The agency is instructed to work with NASA to send "laboratory planes" into the stratosphere to measure the various aerosols already commonly found there. This baseline data will aid any sustained future program to inject additional aerosols.

NOAA is employing customized air force B-57 planes for this purpose, stuffed nose to tail with seventeen different measuring devices. It's a small first step toward a new frontier, perhaps right around the corner, where we mimic nature's volcanoes and reach, humbly or not, for godlike powers.

MARK HARPER, MY pastor in Takoma Park, sat on an old wooden pew at the very front of our church. It was a bright Wednesday morning in late April, and sunlight streamed through the stained glass windows like a cliché.

I entered the one-hundred-year-old stone sanctuary, a block from my house, not knowing what to expect. This was a prearranged meeting. I was guessing that Pastor Mark—fully immersed in the god business—might have some guiding thoughts on the theological implications of geoengineering the planet.

But first, that two-minute walk from my house, at the peak of spring, was like a mini stroll through the glories and miracles of creation itself. More than two weeks had passed since my visit to the rooftop robots at Building 33, and in that time, the oaks in my neighborhood had gone into a state of ecstatic eruption, aided by the third-warmest aggregate temperatures ever recorded for the month of April in this region. Both male and female flowers had opened up by now, blooming on the same tree, the ever-thicker plumes of pollen settling atop emerging acorns that, in turn, five months from now, would tumble to the earth in the billions. The birds, too, continued their push for offspring, with resident robins and northern cardinals building homes in holly bushes and other shrubs. Migrant songbirds, meanwhile, were starting to arrive, including all manner of neotropical warblers (northern parula, yellow-rumped, American redstart) pausing on their way north. The birds hunted insects in the trunks and upper branches of oaks and other hardwoods while their birdsong filtered down from the canopy, a mix of musical peeps and fluttering chirps and soothing whistles as bright as panpipes. The songs mixed perfectly with the explosion of new color along the ground from fresh-blooming flowers: bloodred azaleas, lavender-pink redbuds, violet-blue woodland phlox.

By the time I reached the church, my shoes were adorned with several strands of catkin flowers, sticking to my soles. Though reduced in numbers, the oaks were still cranking out a ridiculous volume of these golden, stringy flowers that clogged gutters and formed rogue tumbleweeds along sidewalks when they fell. I stomped my feet before entering the church. I saw catkins draped from the church's freestanding mini food pantry, too, along the sidewalk—and I brushed them off. I peeked into the pantry's shelves, behind an unlocked glass door, a resource that was free to anyone in need in the neighborhood but was perpetually in need of replenishing.

"The pantry's empty again, Mark," I said by way of greeting as I entered the sanctuary, taking a seat next to my pastor on the front pew. "I noticed it on the way in."

"I know, I know," Mark said, his brow furrowed under a thick head of salt-and-pepper hair. "It's a nonstop battle keeping that pantry stocked."

The congregation of Takoma Park Presbyterian Church had set up the pantry soon after Mark's selection as pastor in 2021. It matched his ministry's emphasis on poverty and social justice issues. But finding a daily supply of noodles and dried beans and canned goods—that was a challenge.

Mark was dressed simply in casual leather shoes and a pullover fleece shirt. He spoke with the sardonic humor of a sixty-year-old minister who'd been around the block. "How can I help you this morning, Mike?" he asked. "We can barely keep the community pantry stocked and keep this church standing, but how can I help you save the planet?"

In truth, he was a dedicated voice on climate issues. In his Earth Day sermon that spring, he declared from the pulpit, "We are waging war on the planet." And on the church website, he wrote: "God longs for us to remove our knee from the throat of the earth so we can all breathe."

The son of a liberal North Carolina preacher, Mark drew inspiration from many groundbreaking activist Christians, including Mississippi civil rights leader Fannie Lou Hamer and Pittsburgh Presbyterian Fred Rogers of children's TV fame. We're all meant to fight ferociously for justice, in other words, but we've got to laugh, too.

Laughter, it turned out, was in no short supply that morning in the sanctuary. It poured in from outdoors, through the stained glass windows, flooding all the pews. To the melody of spring birds and the brightness of new flowers and the hard work of trees, April also brought the migratory song of young children. Their voices floated through the air after months behind winter doors. A dozen kids were running and hopping just a few feet away from Mark and me, students of the preschool next door. They were at recess, chasing one another on the church playground. The scene would have been happily idyllic, reassuring to a tee, if not for the surreal nature of my visit that day. Amid all the new climate worries and flooding issues confronting this church and this neighborhood, I asked Mark if he had ever given thought to the issue of geoengineering.

"What do I think about people playing god with God's creation?" Mark asked. "First, is this really a new issue? Nuclear weapons, the human genome, artificial intelligence. That train left a while ago, didn't it?"

The prospect of geoengineering did not put him in a state of alarm, he said. Mostly, it just made him sad. "After two thousand years, our job now is to be good Christians in a warming world. I think you'll find Jesus, in the Gospels, encourages us to live more lightly on this earth. So I support electric cars. I certainly do. But I support *fewer* cars even more. Isn't that a good approach?"

Above all, Mark agreed, the climate crisis is a human rights violation against poor and vulnerable communities everywhere, here and abroad. And any response must put their welfare first.

Just then, at the front door of the church, there was a loud knock. Mark and I turned around. "Excuse me," said an older woman, peeking her head into the sanctuary, her voice booming. "Do you have any food? The pantry is empty."

Mark jumped to his feet and walked quickly to the woman, who was dressed in a colorful African head wrap and batik shawl. Mark shook her hand. She was in her seventies, out of work, and with no home address, she said. She had been told she could pick up a bag of food here.

"Let me see what we can find in the kitchen," Mark said, and the three of us went downstairs in search of food. After some rattling of pots and the opening of cabinet doors, the verdict was . . . no food. We'd been emptied indoors and out. So Mark started making calls until Gary, the building manager, showed up with a bag of noodles and bread and some lunch meat—and the woman departed, declining the offer of a ride somewhere.

Mark and I kept talking, returning to the pew. He gestured to where the unhoused woman had just stood, wondering how the world's poor, already under epic stress, could possibly manage more heat waves and general climate chaos.

Coincidentally, Mark and I had both lived—at different times—in the Democratic Republic of the Congo in the 1980s, he as part of a church mission, I as a Peace Corps volunteer. I told him about a recent survey I'd come across of diplomats from around the world. It found that leaders from poorer countries—many already facing existential threats from climate change—were more likely to support atmospheric intervention, even with the risks. My own view is that geoengineering is a potential power equalizer, since poorer nations like India and Brazil could plausibly do this on their own. Indeed, Pacific Island nations could theoretically pool their resources for some kind of intervention, as could African states—creating enough diplomatic leverage to force developed countries to give them a full voice in any international decision on whether and how to proceed with atmospheric geoengineering.

"If we have half a chance to save humanity and the

earth, we should consider all ideas, including this one," Mark said, summing up his thoughts. Then he repeated his earlier comment: "It just makes me sad that we've reached this point."

That month, for the first time in his preaching career, Mark was leading a Bible study class on the relatively obscure book of Lamentations from the Old Testament. It is, as the name implies, a long meditative poem on how and why we grieve. It's also full of tragedy and violence. "I don't want to be sad in this world," Mark said. "But it seems to me—the way the world is now—we have to embrace grief. Just submit to it. Make sense of it. Embrace it."

Sitting in that beautiful old sanctuary with Mark, the sunlight pouring in amid waves of children's voices from the playground, I confessed to him I'd been trying for years to outrun grief myself, striving to stay ahead of the sadness of climate change by working and organizing constantly. Every year, it got harder.

"Read Lamentations," he said.

And then, as Mark and I stood from our front-row pew, ending our conversation, I realized something: the laughter outside—it had never stopped. From the preschoolers, four- and five-year-olds, it went on and on the whole time we talked. The children shouted playground commands and talked excitedly and kept on laughing. It ebbed and flowed, that sound, growing louder and then faint, but never dying out completely even for a moment.

6

The End of Fire

May

"Mike! Mike! Your house is on fire! It's on fire, man. Where are you? Three fire trucks are outside your house!"

It was my neighbor Steve, shouting at me through his cell phone on a chilly Thursday morning in November 2007. The panic went straight to my legs. I hung up and began sprinting toward my house, three blocks away from my old office on Eastern Avenue in Takoma Park.

As I ran, I prayed it wasn't true. What could have triggered a fire? Beth was at work. Sasha was at school. Breakfast that morning had not involved the stove or oven.

I turned the corner onto Laurel Avenue, two blocks from my house, and sure enough, there was a plume of smoke in the distance. It rose from the direction of my home. But it was faint. Had my century-old house already burned down?

Still running, I thought about the corn stove. Was that it? For seven winters, I had been heating my home with an ingenious stove that burned corn kernels. The stove sat downstairs in the living room, burning 24–7 in the winter, flames dancing behind a glass door. It was an old Franklin woodstove design, but with a hopper in the back holding seventy pounds of corn. The dried kernels were fed gradually into the flame by an auger.

At the time, this was the best climate solution I could afford for heating my house. The organically fertilized corn absorbed carbon as it grew on a local farm, then released the carbon in winter when burned—a wash. In the shifting chain of fuels that had powered the 7100 block of Willow Avenue over the previous century, corn was the newest arrival.

Sprinting now through tiny downtown Takoma Park, passing shops one block from the flashing fire trucks, I followed a sidewalk lined with city streetlamps. Those old lamps, like guideposts to energy history here, were equipped with ultraefficient LED bulbs today where flickering gas flames had burned a hundred years ago. When I moved here in 1991, many houses—including mine—still had the remnants of old coal chutes built into basement walls. The coal was dumped through the chutes using gravity flow from horse-drawn wagons or trucks, landing beside cast-iron boilers. At my house, the coal boiler delivered steam to iron radiators in every room until the 1940s, when it was converted to oil. That boiler was still in operation when I moved in and installed a modern gas furnace in 1995, heating those same radiators. But six years after that, freaked out by climate change, I paid two farmers from Carroll County, Maryland, to install the corn-burning stove. It was, coincidentally, the same day Saudi Arabian hijackers flew a plane overhead and into the Pentagon on September 11, 2001.

The corn stove at my house served wonderfully and without flaw—until November 8, 2007.

When I reached my house, leaping over fire hoses and sidestepping a crowd of spectators, firefighters stopped me at the top of my driveway. From there, disbelieving, I could peer deep into the interior of my home, into the living room. The fire had burned a sizeable hole in the first-floor wall, and a firefighter was chain-sawing the last of the smoldering wall studs and kicking them outside.

The corn stove had indeed triggered the fire—not because of homeowner error or some design defect. It was the Orkin bug spray guy. He had arrived when we weren't home and began spraying around the outside of the house without giving prior notice. He saw an exterior vent and thought bugs might be living in there and launched petroleum-laced insecticide into the clean-air intake pipe of the corn stove. A flash fire engulfed one wall but did not spread beyond the living room thanks to the quick arrival of the city fire department.

Traumatized by the event and the subsequent months of repair, we reduced our corn use after that, combining it with space heaters while gradually saving money for the newest and final version of heating at our century-old bungalow: modern heat pumps.

The mantra of the climate movement worldwide is straightforward: "Electrify everything." We have to stop, once and for all, lighting things on fire to power our lives. No more combustion inside our homes and businesses for any purpose—heating, cooking, or hot water. No more controlled explosions under the hoods of our cars or at distant power plants. Electricity must power everything—and must itself come from wind farms and solar rooftops and other non-combustion sources. The added benefit: if you don't bring fire into your house, then your home and your planet are much less likely to go up in flames.

Our heat pumps were installed on a cold March day in 2022, and happily, for the first time in 112 years, our house was kept warm without the by-product of smoke. Heat pumps are basically air conditioners that both heat and cool. They use a refrigeration cycle to transfer heat into a building's interior on cold days and cold air inside on hot days. It's amazing technology, powered by electricity, and it works. More heat pumps were sold in 2022 than gas furnaces in the United States. And the price keeps dropping. At the start of the Ukraine war, Joe Biden used his emergency powers to expand US manufacturing of heat pumps so Americans would use less gas during the worldwide shortage. In Europe, the conversion is happening even faster, the goal being to deny Russia the commodity most lucrative to the Putin war machine—exported methane gas.

And here's the thing: we're never going back. Until now, the controlled use of fire by early humans was arguably the most consequential technological achievement of our history. Occurring one million years ago in southern Africa, it spread rapidly to human communities worldwide, with wood the dominant fuel. Wood-burning fires were still the main source of energy for the Anacostan people of the Piscataway tribe who lived, until the 1600s, on or near this land now called Takoma Park. After that, wood fires warmed the scattered farms and lit the camp sites of Union Civil War soldiers who slept not far from here the night before the nearby Battle of Fort Stevens in 1864.

But the rapid paradigm shifts in energy use after the 1860s went in one direction only—toward fuel that was cheaper and easier to use, from wood to coal to oil to methane gas. And with every shift, there was no going back. There are no examples, say, of diesel-powered locomotives switching back to coal-fired versions with the old-fashioned tinder cars in back. Reversals just don't happen. And now that we're making the leap to electrify everything with heat pumps and electric vehicles, there will be no U-turns here either.

There is, in the multiverse of Marvel Comics, a character who, once he gets locked into a full sprint, cannot be stopped by anything or anyone. It's his superpower—the momentum. Ian Chamberlain, the local public works official, brought up this character—named Juggernaut—during our flood tour of the city in March. Ian said the city's job was to slow water down and capture it and hold it long enough during floods that it did not gain, like the Marvel character, the superpower strength of unstoppable destructive momentum.

I think of clean energy the same way. It's a juggernaut on the solution side of the climate equation. It's already unstoppable. Renewable power was vulnerable to delay and commercial failure in its early decades—but no longer. Nothing and no one can halt it now.

Of course, again, there are those who fear that by reflecting sunlight away from the planet, we will create a *moral hazard* that could stall the clean-energy revolution. Critics wonder: Why get off fossil fuels when you can conveniently negate the global warming effects of dirty energy with a little sulfur in the atmosphere?

But to fear this outcome requires ignoring a giant raft of evidence. Passage of the federal Inflation Reduction Act, triggering up to $1.7 trillion in climate and energy investments, is just one part of that evidence. Real-world clues are also all around us.

Near the end of my block, on the DC side, there are two new apartment buildings, just constructed. One is along Laurel Street and the other along Maple Street. Neither building is connected in any way to the gas industry's underground infrastructure. No pipes run to the mains below the streets. From the basement to the lobby to the top floor, these buildings are fully electric. Induction cookstoves fill the kitchens. Heat pumps cover the roofs for hot water and air conditioning. Result: no combustion, no flames, no CO_2 escaping into the atmosphere. And by DC law, half the electricity delivered by the city's grid must come from regional wind and solar generation—with a mandate to reach 100 percent by 2032.

The human species' million-year experiment with fire is drawing to a close. Except for a few limited industrial operations and certain types of aviation, the era will be largely gone in a few decades. It was a great tool, fire. A great breakthrough for *Homo sapiens*.

But getting *off* fire is an even bigger deal, a much greater achievement, one that will have profound benefits for the next million years and beyond for all life on Earth. We're living to see that achievement. It's come a little late and still needs to go faster, but there's no stopping it now.

Meanwhile, at my house, the trauma still lingers. After all these years, I wake up some nights absolutely convinced I smell smoke. I run through every room, looking for flames. But this happens less frequently now, especially since fire itself—from coal, from oil, from gas, from corn—just doesn't exist inside these walls.

Most years, I swear, you can almost hear the leaves bursting from the trees in late April and early May along Willow Avenue. My pin oak, which ends each April with falling catkins and light-green paws of leaf growth, is by mid-May thick with more than half a million fully formed leaves. When it finally comes, when the neighborhood tree canopy finally fills out in the steady march to summer, I'm always startled by how quickly it happens.

Sometimes I wonder: Are the trees growing faster? The gaps in our canopy are more numerous every year, breaking human hearts and ecological rhythms on almost every block. But we still have more trees in Takoma Park than most towns our size. And when those trees turn on in the spring, they seem to sprint, as if trying to make up for the losses. Even in the canopy gaps themselves, the "volunteer" saplings that rise from the ground seem to grow more rapidly than in years past (if they escape the predatory deer).

Turns out, it's not my imagination. The increasing density of CO_2 in the air—after centuries of burning fossil fuels—actually acts as a fertilizer to most of the world's trees and other plants. The additional carbon speeds up photosynthesis. One study found that some species of trees in the Chesapeake region were growing as much as two to four times faster now than in the 1980s.

Which is a good thing from a carbon sequestration perspective. The trees suck carbon out of the air and store it in their bodies for the duration of their lives. Only now, so much warming is occurring that the heat and the weird weather are killing many of the same trees that help store the carbon. And the trees most affected by the extreme weather, at least in my neighborhood, seem to be the older oaks. Which is a double setback.

Oaks are the miracle trees of both urban and rural forests. Wherever they live, more insects (and thus more feeding birds and animals) exist than are found among any other tree genus in North America. Part of it is that so many different moths—and their nutrient-rich caterpillars—have coevolved with oaks, creating a foundational food supply. In his book *The Nature of Oaks*, Douglas Tallamy highlights a study showing how migratory warblers in a New Jersey forest spent threefold more time foraging for insects in oak trees than in pines and six times more than in birch trees. The birds spent very little time at all hunting in any of the other twelve tree types studied. Insect volume explains the behavior.

That same bounty of life exists in urban and suburban yards. Tallamy writes, "There is much going on in your yard that would not be going on if you didn't have one or more oak trees gracing your piece of planet earth." And he told *The New York Times*: "A yard without oaks is a yard meeting only a fraction of its life-support potential."

So the loss of even one oak on a neighborhood block can be a huge blow to the street's ecosystem. On my block of Willow Avenue, we had lost three big oaks in the past four years, and a fourth oak—a giant southern red—was severely damaged by a windstorm in April 2023. It was barely hanging on in the backyard at 7119 Willow. During the rest of 2023, several more oaks would come down on our street thanks to additional strange weather and, we suspect, to a certain hazardous fossil fuel leaking up through the ground below our feet. More on that shortly.

But first the story of May 2023. Most years, again, leaf growth races toward full maturity in May. But a strange and sudden cool snap, after a near-record-warm April, slowed everything down. You could see subtle changes in the leaf growth in the trees. But where you could really see it was in the small patches of ground where human beings try to grow food in Takoma Park.

I'M A LAZY vegetable gardener. I don't get around to planting my main crops—beans, squash, okra—until late April, if I'm lucky. I don't start indoors either. No seedlings below grow lamps in a warm basement. I just walk to the backyard garden, put the seeds in the dirt, and wait.

And that's when it happens, my favorite part of gardening. From tardy beginnings, a mid-May miracle occurs, one worthy of popcorn and a lawn chair. Once the seeds are in the soil, drinking deeply from my watering can, they start—within a week or so—to push tiny stalks into the visible world. Then, with a burst of horizontal growth, the tiny plants throw out what any biologist, if prompted, will tell you are . . . solar panels.

Those first two horizontal leaves start communicating with the sun, ninety million miles away, conducting photosynthesis. And eventually, by early summer, in the colloquial words of Henry David Thoreau, my garden starts to "say beans."

But the month of May 2023 was different. The seeds went in, the tiny stems emerged, the solar panels came out, and then . . . nothing. The plants just sat there, not growing. The problem was obscured sunlight and colder weather. Temperatures ran 2.2 degrees below normal for the month and dipped well into the forties many nights, not so common this time of year here. I gardened in long johns and a fleece jacket in the mornings. It was one of the chilliest Mays anyone could remember across this region. And smoke was a big factor.

While new life was beginning in my tiny garden, old life was going up in flames in western Canada. Unseasonably hot and dry conditions there had mixed with lightning to set two million acres of mature forests ablaze in Alberta and British Columbia that May. Heat rises—and the heat from these record fires was so intense it sent smoke all the way to the jet stream where it traveled to the East Coast. For most of May, a faint haze blanketed the upper sky here, enough smoke to block out enough sunlight to contribute to a cool and dry May in the DC region. Smoke is an aerosol, after all. It impedes sunlight. The photometers on top of Building 33 were very busy that May, recording our region's own mini Mount Pinatubo experience, a reminder of how powerful atmospheric aerosols can be. But this was only a small taste of the wildfire smoke we would see in June, this time thanks to apocalyptic blazes in *eastern* Canada.

America is transitioning to a brand-new energy paradigm without fire. In August 2022, it became statutory US policy. But that policy can't undo, anytime soon, the trend of a planet literally already on fire itself, where nature's greenery turns to smoke and ashes more and more every year.

In the DC region, because of the smoke, the sunsets and sunrises were unspeakably beautiful that May, showy spectacles of deep red and tangerine orange.

I SHARE MY garden with my neighbor Dorothy Lee across the street. Dorothy is a midwife. She's delivered more than eight hundred babies in her career—including many inside the homes of her clients. She says April and May bring lots of babies—nine months after summer vacations—so she's often late planting her garden, too.

Dorothy and her husband, David, are doing their part to end the fossil fuel era. They mimic the renewable ways of our garden not through solar power but through geothermal energy. Ten years ago, they installed a ground-source heat pump. It heats and cools their home by using underground pipes and refrigerants that exploit the difference in the temperature three hundred feet below the surface and the surface itself.

The garden Dorothy and I till is on her side of the street, in the backyard of our neighbors Kathie and Laird Hart. It's a nice arrangement. We cut their grass every few weeks. They let us grow food in their backyard. Our garden now gets more sunlight, sadly, because of the loss of the Millers' massive red oak one yard over, due south.

I was in the garden one very chilly May morning, weeding between rows of struggling beans, when Dorothy came home from work at 8:00 a.m. She had spent the night delivering a healthy baby boy in DC—seven pounds, seven ounces.

But she wasn't happy. "Did you see my tree?" she asked me that morning, bleary-eyed and waving her arms. "The city spray-painted two red dots on it. They're gonna take it down."

Dressed strangely in our cold-weather clothing, Dorothy wearing a scarf, I in fleece and thermals, we walked over to the curb in front of her house. And there it was: Dorothy's beautiful, dying, eighty-year-old willow oak. We were quiet for a moment. The smoke from dead Canadian trees hung in the hazy air, high overhead. Was her tree a victim of climate change, too? we wondered. The bark was rotting and falling off in big clumps on one side. But there were no beetles, and the massive canopy was still in decent shape, mostly leafed out now. It made us mad the city was taking it down so soon. Or maybe we were in denial. I poked the trunk with my finger, and the rotten bark sank several inches below my touch.

Across Willow Avenue that spring of 2023, it wasn't just oaks showing signs of distress. Kurt Lawson, the church volunteer living at Willow and Tulip, watched his Norway spruce, the picture of health, die in just two weeks from a sudden fungus that an arborist said was linked to climate extremes. And on our property, for the first time in thirty years, Beth and I lost a big tree, a silver maple that inexplicably failed to leaf out that spring. It cost $1,200 to take it down.

But now the willow oaks! On Willow Avenue! There were nine of them on the northwest side of the street, giants in a straight line, corner to corner. It was our signature feature, curbside sentries casting uninterrupted shade, boosting the moods and immune systems of all who walked by. It was the most beautiful stretch of mature trees in all of Takoma Park, in my opinion. But Dorothy's would be gone within two weeks, breaking the chain. And the Millers' willow oak, two houses down and part of the same streetside row, was looking sickly, too. It was a great blow to our little block.

A few days later, still mourning her dying tree, Dorothy and I sat on her front porch on another cool evening after another spectacular sunset. She told me it felt surreal some days living through this era of such chaos and deep loss. She increasingly worried about the babies she delivered and their futures.

She told me the story of how she became a midwife. During the early 1980s, she worked first as a hematology/oncology nurse in Pittsburgh, Pennsylvania. She watched many cancer patients pass away over a span of four years. She was there at the very end, at their bedsides, when many of them transitioned from this world. Then one day in rural Kentucky, when Dorothy was on sabbatical, she assisted a nurse midwife during the at-home deliveries of two babies in isolated mountain hollows. She decided this was it. This was the work for her. She wanted to experience the other side of the life continuum and work more independently like her friend Peggy. So she became a midwife.

"I've been at the door—right there—nearly all my career," Dorothy said. "The door where people leave this world and where they arrive. It's the same mysterious door. The great door."

A lonely wind chime dinged softly on her porch as she talked.

"When I deliver a baby, it feels good. But then I come home and, you know, I see this tree is gone and that tree is gone. Some nights, I wake up thinking about my own kids and climate change, and that's it, I don't sleep anymore that night."

The great door. I thought about that image for a long time after sitting with Dorothy that evening. Our fight against climate change—it's at the same door. We're throwing our arms around this planet and pulling back with all our might, standing in the threshold, alarms going off, trying to keep this living world from passing through to the other side, to death and mass extinction and the end of human civilization itself perhaps. All the while, at the same door, amid the convulsions, we're giving birth to something new, something so beautiful, a clean-energy world with fresh breath drawn in, a sparkling cry breaking out, a new life where big trees and all the rest of us can grow old together.

It swings both ways, that door, on my very block, open and shut—the rooftop solar and geothermal energy, the struggling gardens and weary trees, the distant playground laughter of children rising up to a sky full of smoke.

I DON'T KNOW what I expected—but this wasn't it. I was standing in the middle of a busy Takoma Park street, traffic coming both ways, while my friend Barbara Briggs knelt on the asphalt, trying to detect a gas leak.

"Just go *around*!" I shouted to drivers glaring back at me. Then Barbara, with a special measuring gadget in her hand, blurted out the results above the whiz of cars: "Hazard level!" she said. "We've got hazard level two here for explosive risk. Look at these numbers. It's a big leak."

I had not invited Barbara to my neighborhood to get myself blown up. I just wanted to know if there were small leaks of gas on Willow Avenue. It was a growing national problem: invisible, chronic leaks of methane gas, especially in older neighborhoods. Collectively, they constituted a huge climate and public safety risk—and it's another reason why we can't ever reverse course in the switch to clean energy.

So there we were at the end of Willow, near the DC border, standing actually in the middle of the busy cross street, Carroll Avenue, where a so-called gas cap existed. It's a kind of manhole cover used by the local utility, Washington Gas, to access the gas main. And it was literally spewing invisible methane, a heat-trapping gas eighty times more powerful than CO_2 when measured over a twenty-year period.

This odd adventure began a few days earlier when I mentioned to Barbara by phone that two of the willow oaks on my block were really sick.

"Are those trees along the street curb?" she asked. "You

know, it might not be climate change. Gas leaks can kill trees, too. Let me come check it out."

Barbara is an activist with the Sierra Club's Beyond Gas campaign. Equipped with sophisticated handheld gas detectors and some clipboards, she works with concerned residents across the DC region to document gas leaks on streets and in homes. It's "citizen science," she says. She wrote a 2022 report showing how in just twenty-five hours of walking along a selection of DC streets, volunteers documented nearly four hundred active gas leaks spread across all eight city wards, some quite dangerous, all a harm to the climate. On that day in mid-May when Barbara came to my house, she had a hole in her shoe from so much walking and surveying.

Barbara's a devoted Quaker, an avid runner at sixty-four, and a skilled forager of edible plants. She likes to talk with her hands more than anyone I've ever met. To describe how methane harms trees, she took her index fingers and drew a tree crown in the air, then the narrow trunk. Then, with all her fingers fluttering downward, she simulated the roots and then fluttered them back up to show how the toxic gas molecules bubble up and kill the roots.

I was eager to get started that day and pointed to Dorothy's ailing tree, the red dots of death still there, freshly spray-painted on by the city.

"First things first," Barbara said, waving instead to my front door. "Let's start with your kitchen."

Within forty-five minutes, using a different sensor, she determined that our kitchen's gas range and oven—which we were about to replace with an electric system—were borderline poisoning us. We turned on two burners and the oven—and the resultant nitrogen dioxide levels soared, stopping at just two parts per billion shy of what the EPA deems unsafe for sensitive groups. NO_2 and the attendant particulate matter from gas combustion can trigger asthma and other ailments. A 2013 study showed that children in a home with a gas range were *42 percent more likely* to have asthma than with electric cookers. The "clean blue flame" isn't very clean after all.

Next we went outside. We walked south on Willow Avenue, across the street from the big willow oaks. Barbara wanted to get a baseline measurement on this side of the street before crossing over to the trees. Heading toward Carroll Avenue and the DC border, she pulled out her Sensit Gold G2 methane-testing device. It's the same one commonly used by gas companies and firefighters to detect leaks. It's the size of a medium jewelry box with a data display window and a flexible fifteen-inch extension hose and sensor tip.

"Nothing here," Barbara said as we walked down the street. She was waving the sensor across the sidewalk, to the curb, and back again. "I'm seeing zero gas or just one part per million."

We got to the end of the block and were about to turn around and test the trees when Barbara said, "There's a gas cap."

She immediately walked right out into the middle of the street, Carroll Avenue, the cross street, seemingly unaware there were cars coming both ways. I followed her, saying, "Be careful. Be careful." I waved cars around while Barbara's gas tester lit up with warning lights like a Christmas tree—literally. There was a green light showing the sensor was on and functioning properly. Then there was a warning light, bright red, showing that the level of gas leaking from the gas cap was *not* in the low-level category and *not* in the hazard level 1 category. It was at hazard level 2, which meant it was halfway to the level of explosion if any spark were introduced.

At that level, Washington Gas advises you to call their service number to report the leak, which I did that afternoon. I never got a response.

After I managed to escort Barbara off the street and back onto the Willow sidewalk, she said, "That's among the bigger leaks I've seen from gas caps. But even the fire department usually doesn't care. We called them after finding a level-two leak in northeast DC. They sent two trucks, confirmed our test, and said, 'Oh well, you're only halfway to an explosion. Call us if it gets worse.'"

From a safety perspective, the absurdity is plain. Over 4,200 house fires occur per year in the US from avoidable gas leak explosions. But the damage to the climate is just as shocking. Invisibly, all around us and all the time, methane gas is leaking. By weight, it's much lighter than air, so any pinprick hole or crack in the pipe offers escape. Long before it reaches your house, gas has leaked at the original drilling site, at the compressor stations that pump the gas through pipelines, and all along the pipelines themselves. Cumulatively, the US EPA estimates that more than 700,000 tons of methane reaches the atmosphere every year from 630,000 leaks in the US. That's the climate harm equivalent of 4.7 million gas-powered cars. Worldwide, the numbers are just as daunting. And critics say the EPA is probably underestimating the problem. Once released, the methane gas then stays in the atmosphere for a couple of decades, trapping heat at a rate—it's worth repeating—eighty times greater than CO_2. A full 12 percent of greenhouse gas pollution worldwide is linked to "fugitive" methane from human activity, much of it from leaking pipes.

And those leaks kill trees—lots of them.

Barbara has walked over forty miles of streets in the DC region looking for leaks—from gas caps, yes, but also along sidewalks and curbsides for those slower, insidious leaks. Thankfully, my side of Willow Avenue got a clean bill of health prior to the gas cap drama. And later, when we walked the sidewalks along both sides of *lower* Willow Avenue, the 7200 and 7300 blocks, we found essentially nothing.

But that stretch along Dorothy's side of the street on the 7100 block, where the nine willow oaks hugged the curb in a majestic row—that stretch of sidewalk blew Barbara's mind. The gas numbers climbed sharply.

"Is this right?" she said as we walked the sidewalk, under the willow oaks. "I'm getting eight, nine, twelve parts per million here."

She turned the sensor off and back on. Again the high measurements.

"Maybe only once have I seen a leak this high, coming from the ground like this and sustained for a whole block. It's highly unusual."

"Only once?" I said. "So it's bad?"

"Sometimes I'll see a spike of eight or nine ppm in one spot, but not on and on like this block. It's not an explosion risk, but a pipe definitely appears to be leaking down there at a constant rate, and it's finding its way up in lots of cracks in the curb and soil and the sidewalk. Very unusual.

"I don't think you can prove it's killing your trees here, but it certainly doesn't help. It's a problem for any tree trying to grow here."

The chemistry is simple: the leaking gas displaces oxygen in the soil, drying out the ground and eventually suffocating the roots if sustained at a sufficient level. Hundreds of thousands of urban trees likely die this way each year in the US.

"But good luck getting Washington Gas to fix *that*," Barbara said.

Tragically, the trees themselves, growing in proximity to the gas line, could be disturbing the pipeline with their roots. Who knows? What's clear is that trees and methane gas don't mix, from the climate disruptions aboveground to the oxygen-robbing CH_4 in the soil.

Nationwide, activists like Barbara are using these harmful leakage numbers to implement de facto bans on the use of gas in newly constructed buildings in New York State, Washington, DC, parts of Maryland, parts of California, and elsewhere. A more comprehensive ban on the sale of most gas appliances—furnaces and hot water heaters especially—is clearly coming in the next few years in California and many other states.

It can't come too soon. Eight years ago, at an apartment building two miles from my house in Silver Spring, seven people perished—mostly Salvadoran immigrants—in a massive explosion triggered by gas leaking from a failed venting system. Dozens more people were badly burned in the fire's 1,200-degree heat. Again, the *new* apartment buildings at the end of my block in DC, just past the leaking gas cap, don't use this fuel—at all. They're fully electric. The residents there, including those in a building exclusively for low-income senior citizens, don't have to worry about dying in a fiery explosion from a fuel that's also killing the planet and funding wars abroad.

And when the day eventually comes when we turn the gas off forever, everywhere, the trees won't die either. And we can thank people like Barbara Briggs, with holes in their shoes, for that.

A few weeks later in early June, on another hazy day, a tree removal crew came to Willow Avenue to cut down Dorothy's willow oak across the street and haul it away. After eighty years of life, it took eighty-seven minutes for the crew to chain-saw it to the ground and stack it into a truck. I timed them.

A few days later, at dusk, Dorothy and I held a quiet ceremony for the tree. We placed a candle on the ground where even the stump was now gone, ground to sawdust. Per a tradition I learned in Africa, we poured strong spirits—I brought some vodka—onto the ground, marking all four directions of the afterworld: north, south, east, west.

Our tree had passed through the great door, across the threshold, to the other side.

TAKOMA PARK IS home to the first gas station in America to switch entirely to powering electric cars. It's an automotive feat we're quite proud of—but it almost didn't happen. That's because Teresa Doley almost didn't happen. She was born January 31, 2002, at the nearby Washington Adventist Hospital. She was born premature with very weak and tiny lungs. "I don't know if she'll make it through the night," one nurse said.

But she did.

At the time, Teresa's parents, Depeswar and Evaline Doley, were owners of the RS Automotive gas station at Carroll and Grant Avenues in Takoma Park. The couple had struggled for years to become pregnant. They traveled eventually to their native India to pray for a baby at the ashram of Mother Teresa. When they later conceived, they decided to name their daughter after the saint.

Young Teresa loved the outdoors. She grew up among century-old red oaks in the Doleys' front yard on Maple Avenue. She played amid the wondrous collection of bonsai trees her father kept in the backyard. She loved taking hikes along the C&O Canal and beating her dad at stone-skipping contests.

So years later, in 2018, when her father announced over dinner he was closing the gas station forever to focus on repairing cars only, Teresa thought, *Okay, one less source of pollution.* But she didn't actually say anything that night. She was sixteen and usually rolled her eyes and mumbled when her parents spoke. "'Yes, no, ahem,' is mostly how she communicated with us," Depeswar told me later. "She saw us as old and covered in mold."

Until, that is, her dad kept talking that night and said, "Or I could convert the gas station to an electric car–charging station."

Teresa burst into speech after that. She couldn't stop talking. "You've got to do it, Dad!" she said. "You've got to charge Teslas and other electric cars. This is the future. You've got to do it to help the earth."

Depeswar told me all this by phone one day, then invited me over to visit in person inside the lounge where drivers today wait for their cars to charge.

"I'm not sure I would have done this without Teresa's insistence and encouragement," he said in the lounge, a photo of Mahatma Gandhi on the wall while a Chevy Bolt was charging outside, the driver running errands on foot in the neighborhood. "Young people really understand the climate crisis more than older people."

That seems true all across our city where young people flock to the Greta Thunberg–inspired Fridays for the Future events and sketch ecologically oriented poetry on the sidewalks throughout the city. One poem, imprinted on a newly poured sidewalk before the concrete was dry, read:

*From the sky to the earth
and the nature in between
you can see the beauty
in which it is woven
and the space you may weave.*
—Ari Bernstein, ten years old

And it's not just here. Youth climate awareness runs deep nationwide, even among young Republicans, a whopping 67 percent of whom support decarbonizing the economy. It's hard to see a reversal of the clean-energy revolution when Republicans under thirty, the future of the party, see the climate crisis with a crystal clarity lacking among almost all their party elders.

One older Republican, former Maryland governor Larry Hogan, at least pays lip service to the issue. He was still in office in February 2020 when he attended a public celebration of Depeswar's new EV charging station. "He came and was very nice," Depeswar said. "He said very nice things about me and my daughter."

Today, the station looks like a gas station, only with nine-foot-tall electrical charging "pumps" and handheld dispenser plugs under the old service station canopy. The canopy roof itself is covered with solar panels—as is the roof of the three-bay garage and lounge. That building is heated and cooled with heat pumps to boot.

"I'm not an environmental fanatic," Depeswar said. "My daughter didn't push us toward solar and heat pumps. That part was a business decision. We save a lot of money with that."

Drivers at Depeswar's fueling station tend to be apartment dwellers, Uber drivers, or travelers who find the station using EV charging apps on their phones. Local homeowners, myself included with my Hyundai IONIQ 5, tend to charge mostly at home. Walk the streets of Takoma Park and you'll see all manner of front yard charging, from up-to-code driveway chargers to jerry-rigged extension cords running out to street-side cars using duct tape and milk crates to house the dispenser equipment.

Meanwhile—always within earshot, it seems—is that hum, that "faint harmony of angels" as the cars maneuver the local streets.

Most cars at Depeswar's station can get to 80 percent battery charge in thirty minutes or less. The drivers wait in his "state-of-the-art EV lounge" complete with leather seats, complimentary water, and a row of tiny bonsai trees. There's a bakery across the street and Roland's barbershop right next door. "Lots of drivers come over here for haircuts," said head barber Melvin Dawes when I popped in one day. "I'm so fast, I'm finished before their cars are finished."

The state of Maryland helped cover the cost of Depeswar's charging system with a grant, he says. The old gas station was actually losing money. "Now I do a little better than break even each month filling cars with electrons," he says.

Ironically, Depeswar still makes most of his money repairing internal combustion engine cars in his garage. It's a business with a finite future, he knows, with brakes and tires and bodywork the only real repair service in the electric-car future. Clean-energy jobs will boom elsewhere, just not in fixing cars with so few moving parts. Depeswar says he'll probably be retired in ten to fifteen years anyway.

Meanwhile, those gas-fired cars seem increasingly old-fashioned in Depeswar's garage. Oil is drilled in the Gulf of Mexico, then refined into gasoline and transported a thousand miles to get here, then poured into cars and sprayed into engine cylinders for a spark to cause an explosion that powers a piston that makes the wheels turn. That's about 6,000 explosions per minute per running car—or 270 trillion explosions inside 1.5 billion cars worldwide. Every day.

But if there's one fact that spells doom for the Age of Fire—the Pyrocene, as one scholar calls it—it is this: California and the European Union have announced bans on the sale of internal combustion passenger vehicles starting in 2035. Six other US states have adopted California's 2035 ban, including Maryland, and more will likely follow. Meanwhile, General Motors—the icon of American car making—has set a goal of selling only electric vehicles by 2035. And the parent company of Chrysler, Dodge, Fiat, and Jeep has said 100 percent of its sales in Europe and 50 percent of sales in the US will be battery electric vehicles in just five years.

Will these goals be met? Will Trump stop them? We'll see. But the momentum sure suggests that sometime between 2035 and 2040, just around the corner in car years, you'll see almost exclusively electric cars on dealership lots across the country. And then charging stations like Depeswar's will be everywhere and anyone still driving a medieval gas-powered car will eventually need an app on *their* phones to find that rare and old-fashioned and utterly bizarre gasoline filling station.

"Yes, maybe," Depeswar said inside his EV waiting lounge when I mentioned this prediction. "They'll wonder, 'Where will I get my gas?'"

He chuckled, wiped his hands with a mechanic's rag, then went back to stocking motor oil bottles on a shelf inside his garage.

7

Hard to Breathe

June–July

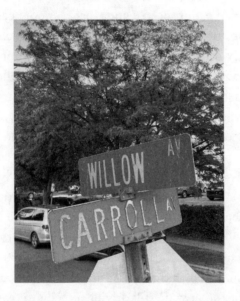

No amount of wind and solar power could stop the planet itself from burning in the year 2023. In June, hundreds of millions of trees died in eastern Canada. Lightning, mostly, brought flames to forests already weakened by drought and high temperatures. By the end of the month, at least two hundred wildfires had spread across Quebec and Ontario.

As the Canadian forests burned, the bodies of the trees turned—in part—to smoke: a combination of water vapor, carbon monoxide, nitrogen oxide, other gases, and particulate matter. The latter is the partially burned remains of the trees themselves—the fine ashes, essentially. Instead of being scattered over land or sea, the ashes from Canada were scattered into the atmosphere, where they traveled via wind to the big cities of the eastern United States—Boston, New York, Philadelphia, DC.

And that's when, going outside, we breathed the trees deep into our lungs. You could smell the forest smoke, filling your nose and throat, as it passed into your body in places like Takoma Park. The smoke dimmed the midday sun and spawned air quality that went from Code Orange to Code Red to Code Purple to, in some places, Code Maroon. I didn't even know there *was* a Code Maroon. At that level, everyone—of all ages—is advised to stay inside. To breathe the smoke meant bringing fine particulate matter not just into your lungs but, for some, into the bloodstream, causing inflammation of the heart and the brain, precursors to cardiac arrest or possible stroke.

Inflamed human bodies—from flaming forests. It had never happened before, smoke like this. Not in the east. At its peak in late June, 120 million Americans were breathing unhealthy air from Maine to North Carolina; from Philadelphia to Chicago. Unlike the smoke in May from western Canadian fires (a high-altitude smoke that lowered our temperatures and prettied our sunsets), this smoke was on the ground and in your face. On June 8, DC had the worst air quality in the world.

With every passing year, it became increasingly clear: Even small amounts of warming brought *big* disruptions to the planet—bigger than most computer models predicted. Some scientists at Goddard, their throats hoarse and eyes burning just walking from their cars to Building 33, were beginning to wonder if the warming so far—about 1.2 degrees Celsius above preindustrial levels—wasn't already too much. The eastern Canadian forests—normally too cool and wet to burn like this—were just the latest example. The fires themselves were part of a terrifying feedback loop. Fossil fuels warm the planet, which triggers forest fires, which converts trees from net carbon absorbers to net carbon emitters, which leads to more warming—and the cycle repeats. The snake was eating its tail.

Through it all, the clean-energy transition continued to grow and do its job. When a sizzling heat dome settled over Texas that same June, making the state hotter than 99 percent of the world, the notoriously shaky electrical grid of Texas was kept functioning thanks to a massive recent growth in wind and solar power plus utility-scale batteries that provided backup power previously nonexistent in the state. This wasn't the PR spin of clean-tech executives. It's how Republican-appointed regulators explained the state's newfound grid stability.

Still, the cycle of warmth continued—and the forests burned.

Ning Zeng was still trying to disrupt that cycle by burying—in low-oxygen conditions—as many dead trees as possible, whether from the sidewalks of Takoma Park or the charred remains of wildfires. But by June, he was still struggling to get the environmental permits he needed from the state of Maryland to properly bury a first batch of five thousand tons of wood from the tree junkyard at Camp Small in Baltimore. To top it off, his wife was begging him to stay indoors for much of June like everyone else, skipping the long jogs that helped him blow off steam from the vexingly slow pace of saving the world.

Just walking the streets of my neighborhood was a surreal experience that June. The smell of smoke came and went, but the haze hung on for weeks, clearing only after a rain or a shift in winds, but always coming back. On many days, health officials advised everyone to wear N95 masks outdoors.

Yet seeing the smoke on the ground was nothing compared to seeing it seven stories up in the air. The first day I went to my high-rise office after the smoke arrived, I couldn't believe it. There, from the edge of Takoma Park, I couldn't see Washington, DC. It had vanished. No National Cathedral. No Washington Monument. No Potomac Valley. No Virginia. I could see maybe eight blocks before the smoky streets vanished into nothing.

What *was* easy to see was an emblem. As the most powerful capital city in the world, Washington had wasted decade after decade, failing in bipartisan fashion to rally the globe while there was still time to fight the moneyed interests of the fossil fuel industry. And now the city itself was literally socked in with unbreathable air from wildfires a thousand miles away and years ahead of schedule on the doomsday calendar.

But then something new happened that summer. Congress, always late, had asked the White House to come up with the beginnings of an actual plan B. In a bill passed quietly in 2022, Congress mandated the development of federal guidelines for research and experimentation related to reflecting sunlight away from the planet. This meant more than just flying NOAA airplanes to gather baseline data in the stratosphere. This might actually lead to a policy blueprint for doing something big. The report was due by midsummer.

So as I stood there at my office window that dystopic June day, it was odd to think that somewhere out there amid all that smoke, lost among the shrouded government buildings, was the White House Office of Science and Technology Policy. And in it, air purifiers no doubt blasting, were staffers putting the final touches on a framework report that, coupled with renewable power, might finally help us see past the smoke.

NING WAS IN China, in Sichuan Province, finishing up a visit with his eighty-nine-year-old mother when his email inbox blew up. His wife was emailing him. So was his daughter. And his grad students. It was June 8, and everyone back home in Maryland was freaking out about smoke for some reason, asking him to write back. Ning went online to check the air quality monitor—the one he had built himself and placed nine years ago on the roof of his University of Maryland academic building—and then he rubbed his eyes.

"I knew it wasn't an error," he told me later. "But the particulate matter was so high—four hundred micrograms per cubic meter—that it was crazy. Our system had never seen anything close to that before."

I was visiting Ning at his campus office in College Park, a seventeen-minute drive from my house. It was early July now, and the worst of the smoke had passed—for the moment. Ning opened his laptop, grabbed a pencil as a pointer, and showed me years of data from the rooftop air monitor two stories above us. A bar graph popped up of daily readings of PM2.5—the finest particulate matter that is one hundred times smaller than the width of a human hair. This is what can get into your lungs and your blood. The bar graph was more or less a flat line of low numbers, day after day, month after month, for years—until June 2023, when tall vertical bars shot up like skyscrapers among tiny buildings. "Normal is maybe five micrograms of PM2.5 per cubic meter of air," Ning said, running his pencil across the data. "Sometimes we get zero. Certainly never more than fifty. Then we got four hundred."

Like the surrounding region, the campus turned into a ghost town, with students fearful to venture out. *No wonder young people reported such high levels of climate anxiety in this country,* I thought. Youth climate activism is up, yes, but so is extreme gloom and depression among people under thirty, polls show. Every generation has its fears. As a Cold War kid in the 1960s, I remember going to bed at night literally wondering if I would wake up again. I imagined myself in a nuclear attack, my body just disappearing, totally atomized. But the mushroom cloud never came, thanks to decades of luck and some focused diplomacy.

But now? The cloud *had* come. The disaster *was* here, smoke and all. Not only could the PM2.5 make you physically sick, but new research shows it can trigger physiological changes in your brain that foster higher rates of depression and anxiety. Mental illness becomes a *physical* consequence of the smoke. A new cascading impact. Meanwhile, campus police grew concerned that if the smog got much worse, security cameras might not pick up acts of crime across the university, triggering more anxiety.

Ning was anxious, too, desperate to get going on his tree burial project. To expedite the permitting process, he had scaled the operation way back to a demonstration project, burying just five thousand tons of the Baltimore trees. He had the necessary paperwork from Cecil County officials (north of Baltimore), but state regulators kept insisting that burying wood waste meant the wood, in fact, was *waste* and he would need a *waste* permit like a landfill. That was a horribly involved process—and totally unnecessary. So Ning modified the project to avoid direct state involvement and require instead only a final federal permit from the Natural Resources Conservation Service. He was hoping to bury the wood no later than November.

"It's tough, man. So tough," he said.

I swear he had more gray hair now—mixed with the black—than when we went to Building 33 last March. His eyebrows were graying, too.

But then, as he often did, Ning flipped a switch, moving from a serious tone and furrowed brow to a therapeutic smile and laughter.

"Oh, wait," he said, "the panda videos! Did I show you the panda videos?"

He closed his laptop and pulled out his phone. His next lab class was in ten minutes, and we spent the time laughing out loud at YouTube videos of pandas from the urban park near his mom's house in China. Sichuan is the provincial home of pandas, and now on Ning's phone, they were rolling head over tail and tackling each other and falling comically from small trees.

It was a brief interlude—"Wait, one more video," Ning said—before we dashed off to his lab class in the same building. The agenda that day involved taking students up to examine the air quality sensors on the roof of the Atlantic Building. This building housed many of the university's earth and space science departments and was famous for the massive supercomputer in the basement—Deep Thought II—that Ning had used to run his past climate models.

Jake Wang, a sixteen-year-old intern from California, was waiting in the lab when we arrived, as was Yasmine Tajeddin, a freshman from Iran. At a large table strewn with screwdrivers, soldering equipment, computer boards, and SIM

A crane removes a 150-year-old red oak from the Miller family property on Willow Avenue in April of 2021.

The author at the White House in September of 2022, to celebrate the Inflation Reduction Act, the strongest climate legislation ever passed worldwide.

The Willow Avenue view, looking northeast, in March of 2023, as Yoshino cherry trees bloom early. A string of willow oaks still dominates one curb as, to the left, a large gap in the tree canopy has emerged after three massive, older oaks have passed away.

Daffodils bloom crazy early in mid-January of 2023 in Takoma Park.

An aerial view of Camp Small in Baltimore, Maryland, of an overwhelmed tree "boneyard." Recent growth in dying trees has led Dr. Ning Zeng of the University of Maryland to propose mass burials to sequester the carbon.

Widespread flood damage in Takoma Park on September 10, 2020.

After the 2020 storm, a 200-foot-long elevated sidewalk was built as a flood berm along the Tulip Avenue side of the author's church, Takoma Park Presbyterian.

A familiar sight in recent years: Workers remove a tree from the hard-hit Lawson-Feasley property (white house) at Willow and Tulip Avenues. Until recently, the property was dominated by large trees.

Dr. Tianle Yuan on the roof of NASA's Building 33 in Greenbelt, Maryland. Sun photometer robots are visible in the background, scanning the sky for planet-cooling aerosols all the way to the upper stratosphere.

A view of Washington, DC, from the author's seventh-floor Maryland office in June of 2023. Canadian wildfire smoke completely obscures the normally visible Washington Monument and National Cathedral.

Depeswar Doley inside the waiting room of his EV charging station. Doley's was the first gas station in America to switch entirely to charging electric cars.

Sierra Club activist Barbara Briggs checks for methane leaks from underground pipes on Willow Avenue. She thinks leakage could be making some trees sick.

A majestic pin oak in the author's backyard, stubbornly fighting weather extremes.

Damage from a violent storm in Takoma Park on July 29, 2023.

Pat Neill stands between Willow and Maple Avenues, next to her colossal willow oak that was fatally damaged during the storm on July 29, 2023.

Lorig Charkoudian, of the Maryland House of Delegates, stands with Jamie DeMarco next to an offshore wind turbine off the coast of Virginia Beach.

PHOTO BY NICOLE SCHROYER

In October of 2023, Dr. Ning Zeng (right) visits with dairy farmer Bill Kilby in Rising Sun, Maryland, to choose a field spot to bury one hundred tons of deceased trees from Baltimore and sequester the carbon for up to one thousand years.

An excavator stacking trees in the grave on the Kilby farm in November of 2023.

An early winter view of the depleted backyard canopy on Willow Avenue looking southwest from Tulip Avenue. By December of 2023, the duration of the "snow drought" had reached a record in parts of the DC region.

A massive, elegant post oak survives on the eastern corner of Willow and Tulip Avenues. Despite the climate extremes, many big trees still hang on. But for how long?

A worker from Casey Trees prepares to plant an oak sapling in the author's front yard just before Christmas of 2023.

cards, they were building air monitoring sensors to be used in wood burial experiments—also known as wood vaults—and other biomass projects from Maryland to West Africa.

"We taught ourselves everything over the years right here in this lab," Ning said. "I learned how to solder right here."

After a few moments, we went up to the roof, climbing narrow, creaky metal stairs until the campus was visible all around us from five stories up. Rooftop pea gravel crunched under our feet. The horizon was hazy but not too bad as the region awaited the next pulse of smoke from Canada.

The rooftop air monitors, meanwhile, the ones Ning's students relied on, appeared to be in good order, recording methane, carbon monoxide, carbon dioxide, particulate matter, and nitrogen oxide.

Calling him "Professor Ning," the students snapped pictures and took notes. Later, I asked Yasmine what climate strangeness was occurring in Iran.

"Dust storms," she said. "They are so bad right now. No one has seen dust storms like this before—and so many."

Jake, from California, simply listed the state's woes: "Droughts, mudslides, fires, atmospheric rivers. Should I go on?"

The speed of it all. The speed. There on the roof, I told Ning that's what the choking wildfire smoke here said loudly to me: speed.

He nodded. "Ten years ago," he said, "I visited my first wildfire site out west, and I realized I had to include this feedback loop much more in my own climate models. The trend was really becoming clear to me and other scientists a decade ago. Now, all over the world, forests are burning. And ice everywhere is melting faster than we expected."

Then Ning mentioned the Camp Small wood storage yard again. The trees there, he said, were continuing to arrive in massive numbers, overwhelming the yard.

"Remember that mound of wood chips we climbed up on in February?" he said. "Remember we stood up there and it seemed like a mountain? Well, that mountain is twice as tall now. It's twice as tall."

THE 103RD ANNUAL Independence Day parade of Takoma Park, Maryland, was scheduled to come down Willow Avenue in just a few days—and everyone on the block was scrambling to get ready. We trimmed hedges and weeded gardens and mowed lawns. The city sent over a street sweeper truck, and workers hung American flags from telephone poles. But before revving up my own electric mower and attacking my scraggly patches of grass to prep for the parade, I did what I always did before mowing: I looked for baby oak trees.

Since 2019, when so many trees began dying and our own giant oak went through an eventual health scare of scattered dead branches, I had set a goal for myself: help my seventy-year-old *Quercus palustris* have a baby in the backyard. The towering oak, still recovering nicely after months of kelp treatment in the soil, was by now lush with the deep-green foliage of high summer. All along the block, shrubs and vegetable gardens and understory trees had turned similarly verdant while up above some nesting birds—robins, mourning doves, song sparrows—had begun their second clutches of the summer.

The problem with my backyard oak, meanwhile, is it had no obvious progeny. Nowhere on the block was there another pin oak. Fearing that the next climate extreme might, god forbid, take it down, I became obsessed with seeing that its gene pool got passed on to a newcomer in this yard.

But being a lazy gardener, I thought it best to let the squirrels or blue jays plant the winning acorn. If a seedling appeared, I decided, I would nurture it and secure its growth.

Those squirrels and jays, however, were efficient eaters—devouring the acorns on the spot or uncovering the countless burial locations the rest of the year. And the ones they forgot? They yielded seedlings that poked their heads above the soil only to be chopped up and chewed to pieces by that marauding purveyor of ecological chaos: deer.

Takoma Park is overrun with deer. When I moved here thirty years ago, there were no deer. Today, you can't go more than a few days without seeing two or three, and sometimes more. They nest amid backyard leaves and hop over fences and cross busy streets. With my own eyes, I've seen a ten-point buck saunter down the sidewalk at midday as if he owned the place.

Deer are also perfect spreaders of Lyme disease, of course—so much so that the primary offending arachnid is named for them: the deer tick (a.k.a. black-legged tick, *Ixodes scapularis*). Just a couple of years after seeing a deer for the first time in my yard, I was sick with Lyme (2008).

It's not their fault, I know. We've eliminated their natural predators and created a suburban "edge" habitat that's perfect for them, where clumps of trees form protective edges beside meadows of green lawns and gardens.

And if there's one food deer seem to love above all, one they seek out and destroy in their daily grazing across my neighborhood, it's the tender green leaves of baby oaks. Across the region, wherever there are deer, our oaks can't reproduce naturally. This creates not just a sterile understory in the urban and suburban landscape—it annihilates succession growth in the vital urban forests of nearby Sligo Creek Park and DC's Rock Creek Park.

I rarely see oak seedlings rise up anywhere in my yard, and when they do, the ungulate stomping and chewing is just too much. The survival rate is literally zero.

Given these struggles and given the loss of our yard's elder silver maple in June, I realized it was way past time to plant a couple of nursery-raised trees on the property—anything to help the neighborhood canopy. Casey Trees, a local nonprofit, was offering free plantings in Takoma Park that year. But what species should I plant? What will last in a climate sprinting toward new extremes?

"A bald cypress," said Keith Howerton, the young arborist with Casey Trees who came by one morning to walk the property. He was dressed in steel-toed boots and an orange safety vest. "A cypress will love your backyard. Conditions are wet and boggy back there a lot, right? Not a problem. The bald cypress grows from Florida all the way up here because it does well in both wet conditions and droughts, and it likes the warmth. It's moving north."

Okay, what about my front yard? It drains better, I told Keith, but still there's so much weird precipitation.

"A swamp white oak," he said. "It does well in wet conditions, too, and it tolerates dryness. It should do well."

Before Keith left, we returned to my backyard one more time, and I gestured to the big pin oak. "And this?" I said. "Is it crazy to plant this here?"

"You can do it," he said. "But you want trees with a good chance of surviving recent trends, right? Your pin oak is a lowland species. It can tolerate wet conditions, with limits, which is why it's probably survived all the recent rain. But it doesn't always do well in periodic drought conditions like the trees I'm recommending."

I went with the ready-for-all-extremes bald cypress and swamp white oak. Climate adaptation right outside my doors. The trees would arrive in the fall.

But I didn't give up my dream of helping my pin oak bring new life into this world. If planting a tree—of any kind—is a universal act of hope, then coaxing a native tree to reproduce in an endangered climate is hope squared.

So before I mowed my lawn that summer day, July 1, prepping for the Independence Day parade, I went through my usual routine. I put on anti-tick spray, then slowly scanned the tall backyard grass and the yard's shrubbed borders for any oak seedlings hiding below. Once again, there were none, despite the thousands of acorns that fell last autumn. So I mowed my lawn.

When I cut the grass, I have to drag a long extension cord with me wherever I go. Not only is it a pain (I really need to upgrade to battery power!), but even if a baby tree somehow escaped my inspection and survived the blades, that cord would likely clothesline it and rip its head off.

As I mowed that afternoon, Beth kept motioning for me to wrap up and come inside. The air quality index was bordering on Code Orange that day due to routine summer smog and lingering Canadian smoke. The cold glass of water tasted great when I finished, and I sat on the back porch steps, surveying my work.

I was about to go inside when there, thirty feet away, in the shade directly below the canopy of my pin oak, was . . . what? I walked over. I blinked. Yes, a baby tree! It was six inches tall. I had mowed right over it and no doubt dragged the electrical cord over it multiple times—and yet there it was. It was leaning to one side, and two of its five delicate leaves had been nicked by the mower blade. But it was clearly not mortally wounded. The twiglike trunk was in good shape.

In my shed, I kept a bottle of synthetic fox urine to spray around the yard—with mixed results—to keep deer away. I pumped an ample dose around the baby tree. The next step would be getting some chicken wire from the garden across the street and building a protective cylindrical fence around this infant.

But the tree seemed fine for now and it was dinnertime, and I went inside for the night.

"MY MOM SAYS you want to interview young people in the neighborhood," said Joanie Lawson, twenty-nine. "Young people who are considering not having children because of climate change? Is that right? Well, that's me."

Joanie had to talk loudly above the sound of bagpipers marching down Willow Avenue. It was July 4, and our city's famous Independence Day parade was in full swing. The local swim team—go, Feet!—had already passed by, as had the Mecca Temple #10 Strollers in their fez hats and white tunics. Now came the MacMillan Pipe Band, all nine pipers blaring with four drummers.

"Yes," I said loudly above the music. "I'd love to talk to you, Joanie."

She had grown up at the corner of Willow and Tulip, and I'd known her since she was a little kid. We were standing now on the front porch of Jim Witkin and Nancy Flickinger at 7127 Willow. They had put out coffee cake and fruit for friends and neighbors watching the parade.

It was hot, already eighty-eight degrees at 11:00 a.m. But none of us knew exactly how hot it was going to get that day around the globe. By midnight Greenwich mean time on July 4, 2023, the planet would set a new record as the hottest day since modern recordkeeping began—17.2 degrees Celsius (62.9 degrees Fahrenheit). The next day would break *that* record. And the next day, that one.

It didn't help on Willow Avenue that Dorothy Lee's big willow oak was now gone, sawed down three weeks earlier right in the middle of the block. A huge and unfamiliar pool of sunlight blasted the street and the parade, making everything hotter. Spectators in lawn chairs and sitting on the curbs avoided the sunny rupture in the canopy, crowding into the remaining shade of the other oaks.

When Jamie Raskin, the parade's grand marshal, walked by, his hand was damp with sweat when I reached out to shake it. A bandanna covered his still-bald pate, but the cancer was gone and the chemo had ended, and he looked great. The whole street erupted in applause as he passed.

Now amid the bagpipes, Joanie and I continued to talk. I had posted a note on a neighborhood LISTSERV requesting interviews with young people for whom climate anxiety was affecting major decisions in their lives, including having kids. Joanie's mom had seen it and passed it on. Now Joanie suggested we meet on Thursday evening, in two days.

Because it might give her more privacy, I suggested we meet at my vegetable garden. Like many millennials, Joanie had moved back with her folks during COVID and was still there. She had been concerned about climate change, it turns out, ever since middle school. Then, when wildfires decimated Australia in 2019, she was so shocked she walked away from a theater career in Philadelphia and enrolled at

Georgetown University. Now, at our chance meeting at the parade, she told me she had just wrapped up a master's degree in emergency and disaster management.

"You can get a master's in that?" I said.

She nodded.

I GOT TO the garden early that evening, across the street, setting up two camp chairs in a lovely spot between the pole beans and a nearby juniper bush. I brought water and a big bowl of summer cherries.

Before Joanie arrived, I also cut a two-by-three-foot section of chicken wire from a supply Dorothy and I kept in the backyard to prevent rabbits from burrowing into our garden. I needed this piece now to keep the deer from eating my vulnerable oak seedling behind my house. I'd been procrastinating foolishly for almost a week and was determined to cap the fragile tree that night.

Joanie arrived in sandals, with her brown hair braided in long strands that fell below her shoulders. We both remarked on how much she and the block had changed since the days when our families would go snow sledding together down the lower Willow Avenue hill in the late 1990s. My son, Sasha, was a toddler then and Joanie was five or six, and the yearly prospect of deep snow was still reasonably common.

Now as the gold light of July began to dim and mockingbirds started their evening song, it was the recent Canadian wildfires that were most on Joanie's mind.

"It just feels like climate change is coming for us now, doesn't it? People who normally don't think about climate seem panicked that they can't go outside and breathe. And people like me, who can't *stop* thinking about climate, we just go into catastrophe mindset.

"I catastrophize a lot about climate change," Joanie said.

I'd never heard that word used as a verb before. Her dark brown eyes narrowed with intensity as she spoke.

Joanie grew up in a neighborhood much more progressive than most of the country, true. But what she had to say that night, her persistent climate worries, can be heard in different forms from young people far and wide, way beyond deep-blue America.

"When I was a kid," she began, "my house was surrounded by a virtual forest of big trees. One year, a giant branch fell from one of those trees. It was almost like a tree itself, and my parents let it stay there for a while like it had fallen in a forest. We called that branch Treelandia. It was as big as the entire yard, and we played there on the branch and had, like, a government and rules for all the kids who came along to play—in Treelandia."

Today, all the big trees are gone around her house, save one red oak. Four giants lost. And now the sunlight pours down.

She continued, "For years after college, I was still working in theater, which requires working day jobs. Eventually, I realized that my day job was taking up all my time and energy, so I started to think about careers I could have that would give me some financial stability and be mission-driven. I was also aware of the rising chaos of climate change and I wanted to do something to help stop it, but at the same time felt so overwhelmed by the scale of the crisis. Then I realized, these disasters we're seeing, they are never going to stop. Even if we mitigate or reverse the impact of climate change, we're going to experience increased disasters. This is my future, disaster after disaster. So if I go into disaster management, my job could both pay for my health insurance and let me help people dealing with the effects of climate change. And that helps my anxiety."

She took a drink of water. She tugged at one ear, adorned with multiple earrings. The first fireflies of the night were starting to blink on all around us.

"I've had anxiety for a long time," she said. "It's individual brain chemistry plus everything going on in the world. Climate dread is my dominant worldview. I struggle with it. One of my favorite podcasts is *How to Survive the End of the World* by Autumn Brown and adrienne maree brown. The anxiety just reached a crisis in grad school when I was finally diagnosed with ADHD. Having that diagnosis, the self-knowledge, helps."

The podcast, she added, examines the apocalyptic nature of our times, yes, but mostly explores uplifting ways of "surviving it with grace, rigor, and curiosity."

She continued, "What really helps me, though, is my job preparing for disasters. I'm working on a project for FEMA where we try to bring together disaster managers and housing officials in communities to prepare for the loss in housing stock from disasters *before* they happen. We know homes are going to be damaged, and the process of repairing or replacing a home is long and hard. That's true for everyone, but especially if you're poor, so what can we do now to get ready to rebuild and recover faster? Working on disasters ahead of time like that actually helps me deal with my disaster anxiety."

I leaned forward in my chair. "So a big part of your day," I said, "is spent visualizing how houses in communities across the country are going to be destroyed by climate change?"

"Pretty much."

"And are you angry about that?" I said.

"What do you mean?"

"You've mentioned dread and anxiety and lost trees and a career visualizing destroyed homes. Do you feel angry about this world that's been handed to you?"

She stared at me. "Who should I get angry at?" she said. "Where should I start? The generation that launched the Industrial Revolution two hundred and fifty years ago? Or people fifty years ago when the warming became super clear? I don't know. I think I rebel against blame. I use plastics, too. I drive a car, and it's not electric. What really upsets me today is injustice. Poverty makes me angry. The best way to prepare for climate impacts is don't make people poor in the first place. Addressing poverty is how we can become a more climate-resilient country and planet."

Joanie's parents, Kurt Lawson and Jill Feasley, are heavily involved in the Presbyterian Church. And while Joanie stopped attending at age thirteen, the themes of so many Sunday school classes seemed conspicuous in her speech: forgiveness, fairness, compassion.

She continued, "I love my parents totally, but when I told my mom a few years ago that I didn't see myself having kids, I think she took it personally. My generation just has a different relationship with parenthood. We think of the suffering that's coming for children. How will we provide medical care in such chaos? And it's not just climate. Will there even be a democracy when my kids finish college?

"I think my mom took it hard because it hit a key part of her identity: as a mother. My dad and I just don't talk about it. He doesn't bring it up. I don't bring it up."

"Are you sad about the possibility of not having kids?" I asked.

She didn't answer directly. There was a pause. "I think I just gave up on that idea a long time ago," she said.

The darkening garden had become a silhouette of shadowy beanstalks and squash leaves and tomato vines. It was the time of day, I knew, when the deer become more active in the neighborhood, moving through the cooling air. A doe was visible just then in Dorothy's backyard, reminding me I needed to go soon to protect my tiny oak.

But there were cherries to eat, and Joanie and I dug into that succulent bowl of purple wonders from southern Pennsylvania.

"What I fear," I said, changing gears and offering one of my own core anxieties, "is food insecurity. Scientists talk about synchronized global breadbasket failure. What if Kansas, Ukraine, central China, and other grain-growing regions experience drought or bad floods at the same time? Maybe it's just a matter of time."

"Maybe," Joanie said. "Maybe. But human beings are not headed for extinction. I don't believe that. The disasters *are* coming, yes. We'll battle those demons. But this won't be an extinction event. We're going to survive in some way or other."

"That seems like small consolation," I said. "Not an extinction event."

"I know," she said softly. "But some days, that's all I got."

The deer I had seen earlier in Dorothy's backyard was now in the driveway. And I could see a fawn with her, too. They were moving toward the street, in the direction of my house.

I excused myself and walked over to grab the cylindrical chicken wire fence. "Joanie," I said, "you want to see something cool?"

We left the garden and moved briskly across Willow Avenue and toward my backyard, me nervously scanning for deer. Thankfully, there were none as we pulled up, in near darkness, to the big pin oak, cicadas roaring in its upper branches. And there on the ground, the seedling.

I told Joanie how I had literally run over it with the lawn mower, and yet here it stood. Before I put the fence around it, Joanie knelt for a closer look, clearly as astonished as I was.

"The world," she said, gazing at the little tree, "is not going to end. It's not going to end."

I explained that if the deer don't eat it, this baby oak could grow for a very long time even in the shade of the parent tree. "The big oak will share nutrients with it through intertwined roots that help with the lack of sunshine. And when the big tree dies one day, the sky will open up for this little, patient tree."

Joanie was getting a text message from a friend just then and was glancing at her phone, but she kept nodding in amazement and gave a thumbs-up and said, "Not going to end."

Then she said, "I gotta go."

We walked the half block toward her house, pausing one last time at the street corner.

"I want you to know," she said, "that despite a lot of the things I've said, I have hope. You have to have hope to go on, even if you don't see having kids. There will still be incredibly beautiful days in the world ahead, full of blue skies even on a changed planet. And there will be amazing people. And if I don't have biological kids, I can still have foster kids. I can totally see being a foster parent. I like that idea."

Now her phone was ringing, and she waved goodbye and walked away. Tired, I went back to my house in the full darkness, pausing to double-check the chicken wire fence around the baby oak. In the cooling night air, I poured water on the tree's tiny trunk from a backyard watering can.

Foster parent, I thought. *I like that, too.*

WHEN THE WHITE HOUSE report on blocking sunlight to fight global warming finally came out that summer, it was a pretty big deal. With climate alarms blaring worldwide, the report proposed a comprehensive and workable framework for studying solar radiation modification in coming years.

What wasn't good about the report was *how* it came out. The White House Office of Science and Technology Policy released it late on a Friday, just ahead of the July 4 holiday—so there was barely any media coverage. I didn't hear about the report at all until July 6, when a friend forwarded a Politico story. Was the topic so controversial it had to be buried on the Friday before a holiday? Or was this just the work of science-minded bureaucrats doing their job with little attention to fanfare?

Whatever the truth, the report itself was good. The opening paragraph, full of wonky weightiness right below the White House emblem, speaks for itself:

"A program of research into the scientific and societal implications of solar radiation modification (SRM) would enable better-informed decisions about the potential risks and benefits of SRM as a component of climate policy," it said. "Such a research program would also help to prepare the United States for possible deployment of SRM by other public or private actors. A research program characterized by transparency and international cooperation would contribute to a broader basis of trust around this issue."

The next paragraph was even more important in my view. It introduced the idea of risk-versus-risk decision-making when crafting future policy:

"The potential risks and benefits to human health and well-being associated with scenarios involving the use of SRM need to be considered relative to the risks and benefits associated with plausible trajectories of ongoing climate change not involving SRM. This 'risk vs. risk' framing, along with cultural, moral, and ethical considerations, would contribute to the necessary context in which policymakers can consider the potential suitability of SRM as a component of climate policy."

In other words, reflecting sunlight—most likely by dispersing sulfur into the stratosphere—has risks. But risks compared to what? Compared to what we're seeing right now? To the fires and record temperatures of 2023? To unbreathable air and climate chaos on my very block and, by extension, every place human beings live on the planet? And it's all going to get worse, guaranteed? How much more climate risk, I wondered, do we have to accept before we consider the risks *and* benefits of solar radiation modification? As the White House report says, and no credible scientist has disputed, "SRM offers the possibility of cooling the planet significantly on a timescale of a few years."

The forty-four-page report, developed with input from ten federal agencies and with public comment, systematically offered guidelines related to research, modeling, data sharing, environmental justice, and "outdoor experimentation." The latter includes the testing of human-dispersed aerosols in limited amounts using balloons or aircraft over small areas of the earth, such as Greenland or Alaska. But none of this, the report made clear, was a substitute for the "foundational" commitments to clean energy and adaptation as responses to climate change.

The report also referenced the somewhat cosmic nature of intervening in the climate by saying "philosophers and social scientists" are needed to help address the uniquely "human dimensions" and the "justice challenges" of climate disruption and intercession.

But disappointingly, the one thing the White House report did not do was actually commit the US to setting up such a research program immediately. This was *guidance* for future research and potential action—not action itself.

But as former US Department of Energy official Shuchi Talati told Politico, "This report . . . signals that the U.S. government is supportive of well-governed research, including outdoor experimentation, which I think is quite significant."

Meanwhile, the European Union that same week announced its support for international talks on whether—and in what way—to manage the sun's rays on a warming planet.

As it turned out, I read the full White House report on geoengineering the same day Joanie Lawson and I met at my garden on that warm summer evening. Sitting in our camp chairs, fireflies incandescent all around us, I described the report to her and gave my view that intensive geoengineering research and careful experimentation were needed as soon as possible. This would provide wider options for global leaders if, say, synchronized global breadbasket failure actually happened.

Her response was simple.

"I don't want to believe what you're saying is true," she said. "That's all I can say. I don't want to believe it."

"I don't either," I told her.

8

In Sickness and in Storm

July–August

On Saturday, July 29, 2023, with almost no warning from the National Weather Service, the sky over Washington, DC, turned a ghastly gray-black at around 4:00 p.m. It had been perfectly sunny an hour earlier. The squall line of thunderstorms forming along the Blue Ridge Mountains came on that fast, heading east, bringing the most violent wind damage seen in the DC region in a decade.

For the previous three days, extreme heat and humidity had dominated the area, spawning a series of smaller thunderstorm bands that loosened the ground—and the tree roots below—with soaking rain. Then on Saturday, the heat soared to ninety-seven degrees, the hottest day of the year so far, just as a cold front approached from Canada and a large "atmospheric disturbance" developed in the upper atmosphere.

I was at home, reading a book on the back porch, the sleeping porch on our second floor, when the storm hit. This porch is not only close to our expansive pin oak in the backyard, it's actually *in* the canopy, with the tree's outer branches surrounding the porch and coming within a few feet of the screened walls. We call it *the treehouse*.

For a few moments that Saturday afternoon, I stood in disbelief, there in the treehouse, as the big oak began convulsing from bursts of wind. The outer branches were alternately being pushed down, bending toward the earth in a strange way, then bouncing back up and being whipped from side to side.

It was a textbook thunderstorm downburst, where surface-level hot air is first sucked up into the sky by the upper-level disturbance, and then a replacement column of cooler air rushes down, slamming into the ground, then traveling horizontally at very high speeds—in this case up to eighty-four miles per hour across the region. To many, it felt like a hurricane. Or a "non-tornado tornado" as one TV weathercaster put it.

In the brief moment before I fled the porch, I heard a loud roaring sound that came not from the east or west but from directly above. I definitely thought it was a tornado descending on the house. But it was just the down-drafting air crashing onto my neighborhood. I grabbed my cat, Macy Gray, and raced to the basement. Thankfully, Beth was away in Pennsylvania visiting her mom.

After what seemed like an eternity but was in fact just fifteen long and anxious minutes, the worst was over. The wind died down. The roaring stopped. Then the sirens started—police, fire trucks, ambulances—all wailing in the distance. I walked outside in the lingering rain. Somehow our big oak was still standing and the house was undamaged. But debris was everywhere—leaves, branches, runaway trash cans. So violent was the downburst of wind that the long, slender seed pods of a nearby catalpa tree had stabbed themselves into the earth like arrows unleashed by a medieval army, standing upright where they fell, *thunk, thunk, thunk*. Between Willow and Maple Avenues, in the backyard of Pat Neill and Wabi Aboudou, a massive branch from a 150-year-old willow oak snapped off and hit the ground with such force it shattered the window of a nearby home on Tulip Avenue.

According to *The Washington Post*, there were over eight hundred calls to 911 that day in DC alone, an apparent record. Thousands of trees in the region weren't as lucky as our pin oak. I walked around and saw fallen trees on Holly and Westmoreland and Cedar Avenues. One tree lost a huge limb that fell on a car on Maple Avenue. I joined neighbors in pulling back branches to confirm no one was inside the vehicle. The street was blocked for three days.

On Willow Avenue, a thirty-year-old tree of heaven fell exactly between the garages of Ashley Flory and her neighbors Rohini Pande and Michael Gordy, damaging both structures. At the downtown gazebo, where buskers play and drum circles gather, a sizable oak crumpled and fell next to the children's playground. Elsewhere in the region, the National Park Service reported 350 trees down along the hard-hit George Washington Memorial Parkway and adjacent areas. Five hundred tons of wood and debris were carted away. Hundreds of homes were damaged across the region, too. One tree fell onto a Virginia house, killing forty-four-year-old Kenneth Allan Lee Jr. while he was taking a shower.

There were scores of injuries but no other fatalities—and much credit goes to the various interagency rapid-response teams that local governments have put together in this era of growing weather extremes. DC's Homeland Security and Emergency Management Agency, for example, worked with road-clearing crews to prioritize streets leading to hospitals. Crews went out immediately to cut sections from fallen tree trunks just wide enough for ambulances to pass through even as the roads were still closed to the public.

For days, getting anywhere in the DC region required a strong dose of patience as other tree removal crews blocked streets with cranes and cherry pickers. And the noise! Storm recovery world is a noisy world. That first week of August, Beth and I would sit on our back porch and just listen—in surround sound: saws at the gazebo taking down an oak; saws on Tulip Avenue cutting up a Bradford pear; saws somewhere in the distance slicing up god knows what on Holly Avenue. It was hard to be a writer that week. I rose early each morning, getting as much work done as possible before the saws cranked up around 9:00 a.m. and concentration became difficult.

But here's the most amazing thing about the storm: how few people actually lost power. Despite flying debris and falling trees, Pepco and Dominion Energy reported that only 225,000 households across the region went dark. It's hard to say for sure, but I'm guessing that had this storm come ten years earlier, three or four times that many people would have lost power. The difference? Since the last really big wind event—the 2012 derecho, when more than a million homes did lose power here—the regional utilities have made big changes.

Pepco, which serves DC and the Maryland suburbs, spent the last decade adapting to the "new normal" by putting more power lines underground where possible. Aboveground, the company installed tree-resistant cables and additional reclosers, which help keep disruptions small and isolated within the grid, not unlike a circuit breaker in a home. In Virginia, Dominion spent years replacing old-fashioned wooden poles with stronger ones made of pressure-treated wood and using fiberglass cross arms instead of the traditional wooden ones.

The result has been an emerging success story in the otherwise difficult world of climate adaptation. Our power on Willow Avenue did not go out on July 29 despite the fallen or damaged trees on almost every block around us. In the 1990s and early 2000s, when extreme weather events were first beginning to spike, we lost power so many times in Takoma Park I eventually bought a seven-thousand-watt gas-powered generator. But I haven't used it in years. Our grid is now effectively hardened against the toughest blows climate change can throw at us in this region—at least so far. Most of the 225,000 people who did lose power on July 29 got it back within a day or so.

But there's one more reason the lights don't go out as often here as they once did, a reason rarely mentioned when newspapers praise Pepco and Dominion. Since 2012, the utilities have preemptively cut down or "fatally pruned" thousands and thousands of big trees that they deemed too close to power lines. Crews with cherry picker trucks and chain saws have rolled down thousands of miles of surface streets—no cul-de-sac too obscure—looking for trunks and branches with any statistical chance of reaching wires in high wind. Today, by utility design, there are dramatically fewer trees *to* fall. Those lost trees, most of them perfectly healthy, are direct victims of climate change, too.

So when you drive the streets and roads of this region now—including Willow Avenue—you'll notice one side is commonly sunnier than the other: the side with aboveground power lines. Small and harmless holly bushes and redbud trees and crape myrtles are frequently planted in place of the big oaks and poplars and sycamores that once stood there. It's a form of self-imposed deforestation.

This is our small triumph over the increasingly dangerous weather. Climate adaptation served bittersweet.

A FEW DAYS after the July 29 storm, my neighbor Dorothy Lee and I went hunting for ticks on Willow Avenue. We were looking for black-legged ticks, the kind that cause so much Lyme disease in our neighborhood. The timing was coincidental. We had planned this hunt weeks ago, but busy schedules got in the way. Now, as I finally crossed the street to Dorothy's house that first Saturday in August, with recovery chain saws still buzzing in the neighborhood, I couldn't help but think again about the unsettling parallel between people and trees here. The trees had to worry about stronger winds and ambrosia beetles attacking their weather-stressed bodies. The people had to worry about ticks bringing spirochete bacteria into theirs. Climate change, in so many cases, was our common stalker.

Dorothy got Lyme disease four years ago from a tick bite in her backyard. She pulled the tick off her body and forgot about it until sharp joint pain developed in her wrists, hips, and knees a month later. There was no rash (fewer than half of people infected ever see one) but an alert doctor diagnosed Dorothy based on her symptoms and prescribed the antibiotic doxycycline. She recovered fully and resumed her busy work as a midwife.

I also got bit in my backyard, in 2008, and also didn't see a rash. But when symptoms set in, my doctor suggested a Lyme test, which came back false-negative. No antibiotics were prescribed at the time, so I missed the crucial benefit of early treatment. I have been sick with the disease ever since. I got reinfected from a second tick bite sometime around 2014, tests finally confirmed, compounding my illness.

So, understandably, when I arrived at Dorothy's house that August day for our unique tick hunt, I came dressed in my ever-cautious outdoor apparel: light-colored pants tucked awkwardly into my socks and a white long-sleeve shirt buttoned at the wrists (to better see any ticks that might get on me). I brought everything we needed to build a "tick drag" on her front porch: a two-by-four-foot section of flannel cloth, a cardboard tube to attach to one end of the cloth, and some rope to drag the whole thing around like a cape. The idea was to pull the white cloth over the grass and leaves and bushes of a yard, checking frequently for dark spots that crawl. Whatever results we found might help, anecdotally at least, educate our neighbors and remind them to protect themselves.

Dorothy was slotting this experiment into her busy Saturday midwife schedule consulting expectant mothers via Zoom, so we had to work quickly. As we cut the rope and measured the cardboard tube, we discussed the recent storm and mourned the fresh loss of neighborhood trees. We were still grieving for her willow oak, too, the big one that was diseased and finally cut down by the city in June. As a gardener, Dorothy now worried about more than just ticks in her yard. For the first time since the Eisenhower administration, sunlight blasted the whole front side of her house. Her shade-loving "understory garden" of dogwoods and wild ginger plants was taking a beating. I noticed the dogwood leaves had turned from green to bright red. "They look sunburned," I said.

"That's exactly what's happening," she said. "The leaves are being burned by the sun. And my poor ginger plants. They're wilting and dying."

And of course, the house itself was much warmer now for the human beings inside, leading to more air conditioner use in a world where we need to use less power. Similar blasts of sunlight were now reaching other Takoma Park homes after the recent storm. At least Dorothy's AC came from the cool temperature of the ground itself below her house via her ten-year-old geothermal system.

When the drag assembly was complete that afternoon, we stepped off the porch, and I made a few initial sweeps in Dorothy's front yard, finding nothing. She then grabbed the device from my hands. "I think they're in the backyard," she said.

The first time she dragged that rectangular flannel across the back edge of her yard, she found something. We each took a knee, there on the lawn, and peered down at the cloth.

"There!" Dorothy said. "A tick! I knew it. The deer are always eating my plants back here."

I reached for my tweezers and plucked the tick from the cloth and held it aloft. It was an adult black-legged tick, all right, a.k.a. deer tick, the kind that carry Lyme. The only thing worse than finding it here and confirming firsthand that the species was still in our midst, all around us, all the time—the only thing worse than that was the tick's behavior once I lifted it. Though pinned between the tweezer tips, its tiny body began noticeably straining to turn toward me, waving its eight legs frantically in my direction. It *smelled* the CO_2 coming from my mouth and was trying with all its might to get to me, to follow that CO_2 stream to the mammalian blood supply that meant survival for it—and more poisonous disease, very likely, for me.

I was pretty sure we'd find ticks that day, but once we snared one, I'll admit, I lost enthusiasm for this project. We had quickly confirmed the obvious: they were here. The plan had been to start at Dorothy's house and eventually check properties up and down the block. But why go on? Dorothy was already getting texts from her clients and needed to get back to her work. I made a few solo sweeps in my own backyard, rapidly finding another tick, this one the even more common lone star tick (which can spread multiple illnesses, including Rocky Mountain spotted fever). I put both ticks in a plastic bag and promptly destroyed them. Then I put the tick drag away.

The storms, the heat waves, the wildfire smoke—these events were intensifying here, making life hard, but they were episodic in nature, coming and going. The ticks and Lyme disease, on the other hand, were always here, unrelenting, a constant through line of our climate story. On block after block, so many residents have either had Lyme disease and recovered from it like Dorothy. Or they still have it and can't get rid of it. Or, like me for many years, they have Lyme but don't know it and suffer through a long series of misdiagnoses.

Ashley Flory at 7116 Willow Avenue got sick, like Dorothy, from a tick bite while gardening in her yard. This one was a lone star tick that gave her not Lyme or spotted fever but the painful hives of alpha-gal syndrome, triggering a yearlong allergy to red meat before she recovered. (Like Dorothy, Ashley also heats and cools her home with an amazing geothermal energy system, by the way.)

And Ashley's twenty-one-year-old child, Audrian, who is nonbinary, is a self-described "tick magnet." They got Lyme as a six-year-old while pulling weeds in the front yard. The disease was treated successfully with antibiotics. Then came a bite from a dog tick, sans Lyme, that grew into a skin ulcer on Audrian's neck for a month when they were eight. A year later, in fifth grade, Audrian came home from an outdoor school trip covered with over eighty black-legged tick nymphs the size of poppy seeds. There were twenty-three behind one ear alone. With a knife tip and tweezers, a babysitter painstakingly removed all the ticks before any Lyme bacteria could be transmitted.

But the true scale of the problem—in Takoma Park and beyond—becomes clear only when you consider this: the Centers for Disease Control and Prevention estimates that for every reported case of Lyme, *there are ten that go unreported*. The math quickly becomes mind-numbing. At least six people at my church—that I know of—have had cases of Lyme documented by doctors. And the disease has run like a brush fire through the local Boy Scout troop, affecting both parents and boys over the years. In 2021, an Eagle Scout from Troop 33 received a late diagnosis and was bedridden for two years, delaying college and requiring intense antibiotic treatment. He asked not to be named in this book, fearing future employers might not hire a young man with such a complicated disease.

My sister in Georgia has battled chronic Lyme. And three of the board members of my nonprofit, CCAN, have had Lyme. One, Terry Ellen from Baltimore, has seen the disease devastate his whole family with debilitating sickness that continues today.

My family, my block, my church, my job, my volunteer

world—Lyme runs through all of them. Thirty years ago, I knew no one with Lyme in any realm of my life.

In 2015, I was helped in my own illness by Dr. Joseph Jemsek, a noted "Lyme-literate" doctor in DC. With a mix of antibiotics, herbal supplements, and experimental antifungal medicines, Jemsek helped stabilize and then reverse my descent over a multiyear period. I continue to treat the residual symptoms—intermittent flu feelings, periodic brain fog, and muscle stiffness (still no picnic)—with acupuncture and herbal medicines.

Climate change did not create this awful sickness. Lyme was found in the body of the 5,300-year-old "Iceman" famously uncovered by retreating Italian glaciers in 1991. But global warming is a main driver of its spread today. Milder winters mean more ticks survive each year and are active—moving, biting, reproducing—whenever the temperature is above forty degrees. They also thrive in the damp and humid conditions now intensifying across much of the eastern United States, certainly in the DC region. Proliferating deer and white-footed mice also help spread the disease.

So strong is the link between warming and Lyme disease that, starting in 2014, the US Environmental Protection Agency began using the disease as one of several indicators of intensifying climate change. Other EPA indicators include the annual acreage burned by US wildfires, the water levels of the Great Lakes, and seasonal household energy use.

So the explosion in annual US Lyme cases—from an estimated one hundred thousand in 1992 to five hundred thousand in 2022—speaks volumes about warmth in places like Maryland, which has the eighth-highest Lyme infection rate in America. Nationwide, annual infections now outpace cases of breast cancer and HIV combined.

The good news—and there is good news—is that an apparently effective and safe vaccine, called VLA15, is in late clinical trials and may be available as early as 2026 for worldwide use. Equally encouraging are research and development of a body scan that can detect the presence of persistent Lyme bacteria in patients like me. Tests with lab animals have shown for years that the Lyme bacteria—*Borrelia burgdorferi*—can survive even intense antibiotic treatment, especially after late diagnosis. In other words, long Lyme—like long COVID—is a medical fact.

But none of these prevention and diagnostic breakthroughs will slow the spread of ticks themselves as long as the planet continues to warm. And those ticks, worldwide, carry and spread to humans no fewer than twenty-two different pathogens, some of which—an Ebola-like virus in Europe and the nasty *Babesia microti* bacteria here in the US—are as bad as or worse than Lyme.

So it was understandable that on that hot August afternoon with Dorothy, hunting for ticks, I experienced a personal form of intense tick PTSD. I'd spent years avoiding these beasts at all costs and waking up, startled, from tick nightmares. On that August day, I put away the white flannel tick cloth and abandoned plans to test every yard on the block. I told all my neighbors that the device was available and free for their use.

No one took me up on it.

9

Solutions

August–September

I was driving on the DC Beltway one late-summer day when Ning Zeng, my passenger, began contemplating an act of civil disobedience. We were coming back from visiting the sad remains of the oldest white oak tree in Montgomery County, Maryland. There were scientific tools in the back seat that Ning had just used to analyze the tree's trunk. Now he was tired—in so many ways—and he just couldn't hold back anymore.

"If the government," Ning said, "tells me that doing something to solve global warming is a crime, then do I have a moral obligation to commit that crime anyway? Should I do the right thing even if they say it's wrong? I keep asking myself that question."

He was referring to his tree burial project, of course. The plan to place five thousand tons of trees underground on a farm in northern Maryland, showing the world a new way to sequester carbon, was still bogged down in government red tape. And Ning's company, Carbon Lockdown, was almost completely out of money. Ning was taking a sabbatical from teaching this coming fall to focus only on getting the project off the ground. He had begun dipping into his retirement savings, cashing mutual funds, to pay for core operations. Just moments earlier, when we were outdoors in the ninety-five-degree heat and Ning was measuring the massive oak tree trunk, the perspiration covering his face made him look almost like he was crying. Those rivulets on his cheeks—were they tears? It's how he felt on the inside, I knew. Ning was as low as I'd ever seen him.

He had texted me two days earlier, asking if I wanted to visit the Linden Oak in Bethesda, Maryland. *The Washington Post* had just run a story saying the tree, after three hundred years of life, had been cut down. The bottom fifteen feet of the trunk were left standing as a kind of memorial—and hundreds of locals continued to pay their respects throughout the summer.

When we arrived at the spot, just off the Beltway and near a noisy Metro rail line, a pair of mourners were picking up pieces of tree bark as souvenirs. And no wonder. The remaining trunk of this white oak—*Quercus alba*—was hard to comprehend, it was so massive. The tree was born before George Washington. Two-plus centuries later, in 1973, community protestors forced engineers to spend an extra $2 million routing the new DC subway system around the already gigantic tree.

"Here, hold this," Ning said when we reached the trunk. He handed me one end of an eighteen-foot measuring tape. He took the other end and walked around the trunk until the tape . . . ran out! Ning was two feet short of reaching me. "What?" he said. "A twenty-foot circumference? Unbelievable."

He calculated the tree's diameter using *pi* in the equation. The tree was seven feet thick. We just stood there for a moment. Seven feet. His entire life, Ning said, he'd seen only a handful of trees this big. One was the old red bean tree outside his grandfather's village house in China. It was so big, with a hollowed-out passageway at the base of the tree, that Ning and his brother could literally run through the trunk as kids. They could run through the tree.

Ning pulled out a drilling device and, with both hands, began twisting it into the tree's trunk. He was taking a core sample. He would study the size of the tree rings and other features later in his lab. Locals say the tree went into a noticeable decline in the early 2000s. Then, in 2020, as mass mortality continued to spread among old white oaks across the region, the tree lost a giant branch and went into final descent.

Back in the car, Ning did more calculations using pen and paper. He determined the tree's stored carbon when it died was equivalent to as much as one hundred tons of CO_2. That's roughly what twenty gas cars produce in a year, a small parking lot. Toward the end of its life, the tree was probably sequestering another two hundred pounds of CO_2 per year from the atmosphere. That was all over now. Most of the stored carbon in the wood would wind up in a landfill or be mulched—the decomposing fiber venting CO_2 and methane for decades. Only a small part of urban trees like this are suitable for lumber—the rest rapidly lost in the fight against global warming.

Calculations like these always put Ning in an agitated state. The more trees died in this region, the more he felt the moral urgency of his mission to bury the wood, storing the carbon away for a thousand years or more. The good news was that a prominent Swedish investment firm—Kinnevik—had just "purchased" the first thousand tons of stored carbon from Ning's proposed underground wood vault in Cecil County, Maryland. It was part of an international carbon-offset market, which, when properly operated, could allow green-minded purchasers to offset their hard-to-avoid carbon emissions—like the commercial air travel of Kinnevik employees. The company's purchase was a historic validation of Ning's work. It was the first purchase of sequestered carbon in the form of buried wood on the globe. Eventually, in a fully wind- and solar-powered world, there *will* be no new emissions of carbon to offset. At that point, governments and companies will still need to pay to sequester the excess "old" CO_2 already released into the atmosphere, exploiting the bodies of both living and deceased trees for many centuries to help make this happen.

On paper, Ning had until June 1, 2024, to bury the wood and trigger the Kinnevik purchase. But in reality, winter and spring rains meant he needed to bury the trees by mid-November of the current year, 2023, with all the CO_2- and methane-monitoring equipment mounted above- and below-ground. That was just three months away. After that, rain would make digging difficult and potentially cause the long, dirt walls of the burial trench to collapse.

Three months. And the final county and state permits Ning needed weren't even close to being granted. He had appealed directly to the staff of Maryland governor Wes Moore, hoping the governor might personally intervene. He was turned down.

So on that day on the Beltway, coming back from the Linden Oak, Ning could feel his sequestration project slipping away for another year. And after the climate extremes of 2023—bathtub-warm ocean water in stretches across the globe, land-based catastrophes on every continent—he knew the world didn't have one more year to waste.

So Ning started talking about breaking the law.

"If I don't get the permits," he said as we crawled through Beltway traffic, "then it's technically a crime for me to bury the trees. But if I don't bury the trees, I'm committing a crime against the planet. The earth could lose the chance to sequester billions of tons of carbon per year in the future. That's huge. If we don't start now, we may never get there."

He put down the pen and paper he used to calculate the carbon stored in the Linden Oak, a tree older—incidentally—than the fossil fuel era itself and the 250-year-old industrial age built upon carbon pollution.

"Carbon is like a river," Ning said to me, the scientist in him fully coming out now. "It's always flowing somewhere, in the form of gaseous CO_2 in the atmosphere, trees on land, carbonate ions in the ocean, and other forms. Now we know we can divert that flow into the ground as wood. But if you wait ten or twenty years to start, you've missed ten or twenty years of the carbon river flow that you can never get back. It flows into the sky instead. If we don't start diverting the wood, we could forever lose the chance to sequester twenty or thirty or forty billion tons of carbon over the next twenty years. That's just huge for the climate.

"So," he said, "I've spoken to the farmer in Cecil County—Mr. Kilby—and we've made up our minds."

The perspiration on Ning's face had disappeared by now. His cheeks were dry. He folded his arms and turned toward me from the passenger seat. He was wearing a T-shirt that day that said—and I'm not kidding—WHERE THERE IS A DREAM, THERE IS HOPE.

"We've decided we're going to do it no matter what. If we don't get the permits, we're going to bury the trees anyway. We just can't wait any longer."

For the record, months later, he recanted those concluding words spoken in the car that day. He really didn't mean it, he said. And events would later make the question moot. But he sure sounded serious during that low-point drive along the Beltway in the summer heat.

IN MY NEIGHBORHOOD, the sadness of dying trees is often blunted, at least partly, by new opportunities to fight climate change. When trees come down in Takoma Park, solar panels regularly go up. And vegetable gardens happen.

I rooted every day for a breakthrough in Ning's project to help redirect the global CO_2 river. I also noticed every day smaller tributaries of that same carbon stream flowing and stopping and re-forming on Willow Avenue and surrounding streets.

One day, from my seventh-floor office window a block from my house, I watched an acrobatic chain saw operator on Maple Avenue float forty feet off the ground, hanging from a crane's cable, his body the only cargo. Dangling this way, chain saw roaring, he helped take down a mountainous oak next to an old apartment building. Within a month, that building's flat and once-shady roof was covered with a five-kilowatt solar array. This happens over and over again here.

And when the old willow oak on Willow Avenue died that year, Dorothy and I decided to expand our modest gardening empire to a spot right next to where the tree had stood. The space between the sidewalk and the street is technically city property, but we'd noticed a trend of street-side farming on some nearby blocks. So in mid-September, we ordered two metal raised-bed containers online and began dreaming of fall spinach and garlic bulbs planted in November—right on the curb—for next year's pasta.

In the meantime, as summer turned to autumn, recovery from the July 29 windstorm dragged on in the neighborhood. On September 14, one of the biggest willow oaks in the entire city came down just behind Dorothy's house, in the backyard of her neighbors Pat and Wabi. Its trunk had been fatally damaged by the eighty-mile-an-hour July gusts. A crane several stories tall was brought in for the job. Between Willow and Maple Avenues now, the whole backyard tree canopy was just disappearing. And the Presbyterian church members, eyeing the loss from their downslope, flood-prone property on Tulip Avenue, grew increasingly nervous about the faster runoff of rainfall pouring their way.

By September, worldwide, enough data on extreme heat and anomalous weather had poured into the National Oceanic and Atmospheric Administration that the agency declared the year 2023 on track to become the warmest year ever recorded. The year had a 93.42 percent chance, the agency said, of beating out 2016, the previous record holder. July had already become the hottest *month* on record—by far. And in the Atlantic, extremely warm water was bleaching coral reefs and giving rapid formation to hurricanes. In Antarctica, sea ice was at the lowest level ever seen for that pole's winter months.

Farmers, in particular, have much to lose in this warmer world. Some climate models show Kansas turning into a virtual scrub desert if the warming rises several more degrees above preindustrial levels. Already, many wine growers in California are moving their vineyards to higher elevations, and some farmers along Maryland's Chesapeake Bay are losing nearly all their cropland to saltwater intrusion from sea-level rise. Amid these climate shifts, Big Ag is investing heavily in resilient seed varieties and in drones and satellite data to create "precision planting and harvesting." But no amount of behavior change and new technology is going to forestall a big subtraction in agricultural yields if temperatures maintain their rapid rise.

So for me, it wasn't just a hobby that led me to grow more vegetables in this newly sunny spot on Willow Avenue. The success of any realistic climate adaptation plan for this country must include an emphasis on distributed agriculture. That's where as many people as possible grow as much food as possible near their homes as a buffer against extreme climate events affecting farms worldwide. Breadbasket collapse will bring less harm if every household produces at least a few baskets of food themselves.

My friend Denny May, a climate activist and gardener who lived for years on lower Willow Avenue, saw this coming more than a decade ago. When he retired from teaching English composition to mostly immigrant kids at a local community college, he started a second career as an urban farmer. He bought a pickup truck, got certified as a master composter, and began recruiting people like me. I had gardened for years in my own backyard spot before shade from the pin oak put me out of business. So in 2014, Denny and I asked Laird and Kathie Hart across the street if we could garden their sunny backyard patch in exchange for lawn mowing. They agreed, and Denny and I shared the plot until he moved to Vermont in 2021 and Dorothy took over his share of the garden. Before he left, Denny had established a half dozen small plots like this across Takoma Park and DC, driving around in his pickup truck, dispensing mulch, and harvesting squash and boxes of raspberries—the urban farmer.

The vision here is not quixotic. During World War II, when the need for soldiers depleted many farms, Americans grew a whopping 40 percent of their vegetables in victory gardens set up in backyards, vacant lots, and even windowsills. Today, home gardening is exploding again. The COVID pandemic alone got tons of people tilling outdoors in urban and suburban areas, experiencing the outdoor air and sense of community that gardening often brings. The year 2021 alone brought in 18.3 million new gardeners nationwide, according to the National Gardening Association, a one-year rise of more than 25 percent. Demand for seeds led some suppliers to run out. And if my town is any guide, additional numbers of people are gardening because big trees are falling in many urban areas, closing one door and opening another. Even when young replacement trees are planted nearby, a must, those gardens offer years and years of productivity before shade has its way.

Since 2020, new food gardens have popped up in Takoma Park outside my church and next to city hall and in front yards on countless blocks. No wonder *The Old Farmer's Almanac* is displayed next to the cash register at our downtown Ace Hardware. I get my annual copy to learn, among other things, the best moon phases for planting seeds (it's a real thing). Publishers of the almanac say the 2.5 million copies sold nationwide in 2022—a record—reflect both the rising popularity of gardening and the nation's increasingly anomalous weather. More and more people need help understanding the new normal weather patterns and so turn to the 231-year-old publication's combined scientific and "secret" methods for predicting regional weather.

Meanwhile, the question looms: How will our traditional farms, out there in rural America, survive when so much of their land continues to grow too dry, too wet, or too hot to raise conventional crops? One answer is obvious: clean energy. Many farmers can hang on to their dinged-up farms by harvesting wind and solar power as supplemental "crops" and then shipping that energy bounty to all us new food growers in the city. I love the thought of rural wind turbines and rye fields coexisting to push methane-spewing gas plants off the grid; of backcountry beehives and pollinator-friendly flowers surrounding solar panels that shut down coal use.

On a Saturday in late September, the first day of fall, Dorothy and I finally began building our new garden there along the curb of Willow Avenue. We broke out the wrenches and screwdrivers to assemble a pair of eighteen-square-foot metal-walled beds. We filled them with mulch and with a compost made from grass and kitchen scraps, then added a splurge of organic garden soil purchased from a local landscaper.

To the left of our beds, ten feet away, the city would soon plant a new willow oak tree near the same spot where the old one was lost in June. Our garden would have a good twenty years before the shade of that new tree became too much—*if* the tree survives two decades amid the ever-shifting rains and haywire temperatures of this changing block. As we gardened, Dorothy and I could feel the presence of the previous tree, the old-timer. It wasn't there physically to give us permission to dig, but we sensed the unspoken blessing in this circle of space it had sheltered for so long. We feel it every day.

And as the month of September drew to a close, I began to wonder when we were going to get one of those bright and grand and breezy days the month is known for around here—those annual days when a million windy voices from a million shaking tree leaves tell us autumn is arriving; those days when we humans have to raise our voices a bit to be heard on sidewalks and over backyard fences.

And then I realized: those days had already come and gone. On September 15 and 18, my journal shows, it was pretty darn breezy. But the days felt like mere warm-ups for something bigger that never came, something windier with fuller sound. In reality, there were just fewer trees; fewer than even a year ago. The oldest willow oak on the street, right across from my house, was gone. And behind Dorothy's house and all the homes on that side of the street, the canopy was just decimated. So the annual September swishing sound on the block was more like a few jazz ensembles competing with one another, with some strong solos over here and over there against a background of softer leafy percussions. An octet of big willow oaks still lined the street and was loud, to be sure. But that fuller, orchestral sound of wind blowing through a mature urban forest was fading.

FOR YEARS, I'M ashamed to say, I stood by and watched as thick vines of English ivy entombed a lovely American elm in the vacant lot right behind my house. Year after year, it went on. The ivy spread through the canopy and I didn't connect the dots. When I finally hopped the fence and began cutting the vines in a delayed panic, it was too late.

As if the trees on Willow Avenue don't have enough to worry about, there's this: a foreign legion of botanical killers is always scheming to smother them and strangle them and pull them to the ground. It's hard to describe, in fact, the full variety and the naked aggression of the invasive vines attacking trees across this region. It's probably the same where you live, if you look around.

Chinese wisteria, winter creeper, porcelain berry, Japanese honeysuckle—these are some of the top killers on the East Coast. But the worst, in my neighborhood, is English ivy.

I've seen some of our tallest trees completely covered with the dark, green, five-pointed leaves of this ivy. The trunks are suffocated in a tightening girdle of crisscrossing vines—sometimes arm-thick—that can add hundreds of pounds of extra weight to trees already struggling to stand up against higher winds. Absent intervention, the vines just take over, prospering wildly until, inevitably, they kill their stoical host.

Invasive tree vines are not a new problem. They've been around for generations. What *is* new is that the vines are growing faster now. Studies confirm this, showing noxious invasive vines—like most plants—are stimulated by the heightened CO_2 in the air. In my neighborhood, they simply outcompete most native plants while, to boot, rising temperatures extend their growing season.

Thirty years ago when I first moved here, a tree dying from invasive vines was a relatively rare occurrence. Now it's a constant threat pretty much everywhere.

In March of 2020, I couldn't take it any longer. Restless from the COVID shutdown, Beth and I had begun taking longer and longer morning walks in the neighborhood. That's when, for the first time, the full problem became apparent to us. Vine-affected trees were on almost every block.

But what could be done? The city government of Takoma Park was all but ignoring the problem, so I searched the internet. I hoped to find a model strategy from some other town or county—a strategy where vines were eradicated across a jurisdiction using, I imagined, volunteer muscle or a safe organic herbicide or both. Perhaps the US Forest Service had grants and guidelines for city residents wanting to tackle the problem.

But after hours on the computer, searching not just in the US but worldwide, here's what I found: nothing. On the entire planet, there wasn't a single success story that could serve as a model for a city like mine. Dozens and dozens of studies were out there documenting the severity of the invasive vines problem and calling for more research. Countless chat rooms were filled with the complaints and invectives of gardeners and environmentalists and public works officials. But no one, apparently, had figured out a safe and practical way to comprehensively solve the problem of invasive vines in any given town or county.

So I called my friend Jesse Buff, a Takoma Park resident and natural resources specialist with experience fighting invasive plants in parts of nearby Rock Creek Park. There were plenty of YouTube videos out there on how to safely remove vines from individual trees, Jesse reminded me. And there were various Weed Warrior volunteer groups attacking patches of trees here and there. But he, too, had never heard of a successful, comprehensive case study involving the coordinated eradication of vines in a given jurisdiction.

Crazy as it seemed, we were going to have to be pioneers on this one.

Step one, I realized, was getting a firm grip on the scale of the problem. How many trees in our town were at risk of dying from invasive vines? And where, exactly, were they located? From these baselines, we could measure our progress.

So on February 1, 2021, wearing a heavy winter coat and a pair Oboz hiking boots, Jesse took off on the world's first-ever assignment to find and plot every tree clobbered by invasive vines within a city's boundaries. Armed with a map and a smartphone app he helped design, he planned to walk all thirty-six miles of streets and roads in Takoma Park plus another ten miles of park trails. The wintertime absence of deciduous tree leaves meant he could see, even in forested parks and across distant backyards, the evergreen leaves of the most common invasive vines.

The results? Over a monthlong period, Jesse documented a staggering 4,850 trees in Takoma Park that were at risk of dying from invasive vines within the next five to seven years. This was out of roughly 38,000 total trees in the city. Some of the trees were giant canopy trees that the city could *not* afford to lose. The large majority, however, were younger understory trees that represented the future of the city's tree life. When they die, they don't all make it into the city's official mortality count. And absent some type of intervention, they *would* die.

So we intervened. Every Saturday at 8:00 a.m. for fourteen months, volunteers gathered in the parking lot of our local grocery store, a friendly food cooperative on Ethan Allen Avenue. Jesse and I decided against using herbicides for this project, given the extensive training and health precautions involved. Instead, we handed out free pruning saws and garden clippers and gave a brief training on the simple method for liberating trees: cut the vines at chest height, careful not to harm the bark. Cut again at the base of the tree and pull the attached roots as far away from the trunk as you can.

Then we handed each volunteer a list of addresses from Jesse's survey.

For fourteen months, the volunteers kept coming—and coming. Seventy-two-year-old Elizabeth Thornhill arrived many Saturdays on her Bridgestone bicycle with a basket on the handlebars to carry her tools. Nearly a quarter of the volunteers were high school kids looking to satisfy community service hours ahead of graduation. Pretty much everyone grew instantly addicted to the sheer fun and intense meaning of what we were doing.

One volunteer said this: "It's the purest form of instant gratification. You sit at home feeling powerless against COVID and climate change and war, wondering how you could possibly do anything to help the planet. Then you come here and volunteer, and by 11:00 a.m., you've saved ten trees. Those trees were choking and dying, and in a few minutes, you snip a few vines and you step back, and those trees are saved."

That volunteer was David Olson, a Lincoln Avenue resident whose Minnesota roots kept him out there cutting vines even on the coldest Saturdays. When the project ended, he had personally saved 453 trees.

At the start, we focused on trees in priority Takoma Park communities, fanning out around the local senior citizen facility—Victory Tower—and the lower-income apartments on lower Maple Avenue, where landlords commonly fail to keep up with invasive vines. Then we turned to parks and businesses and the yards of private residences.

Once we hit four thousand trees, there was little more we could do. The remaining trees were on abandoned properties with no access or on the private lots of grumpy homeowners who denied permission. Still, the project was a huge success. It was also, admittedly, temporary. We would have to cut the resurgent vines all over again five years from now.

But this first-in-the-world and ridiculously simple method really worked: One, conduct a full tree survey. Two, set up a system of regular volunteer hours. Three, have fun. It's now being replicated in nearby Hyattsville, Maryland, and, soon, across large parts of DC.

Along the way, of course, we made a lot of friends, conversing over backyard fences, educating folks in parking lots. We also observed, again and again, the often mystical, emotional relationship people have with trees.

David Olson, the star volunteer, one day knocked on the door of a modest, two-story home in northern Takoma Park. When an elderly woman answered, he told her the three tulip poplars in her backyard were in trouble. Those English ivy vines were killing them, he said—but he'd be happy to cut them, and the trees would recover nicely. She agreed and escorted him to the backyard, where he went to work.

It was an exceptionally hot and humid late-summer day, and David's T-shirt was already steeped in sweat as he took on the first tree. Curiously, the woman, about eighty, stayed out in the heat with him. Homeowners normally went back inside as he worked. But she stood in the dappled shade of some hedges, watching him intently from about fifty feet away. It was a little strange, but she insisted on staying.

David kept working, clipping and sawing vines until he finished the first two trees. Then he turned to the final tulip poplar, the biggest, probably sixty years old.

And that's when the woman emerged quickly from the nearby shrubs and walked toward him and stood next to him. She looked up at the tree's crown, far above, then back to the base of the trunk where David was kneeling.

"My daughter," the woman said, "she died."

David put down his pruning saw and turned to her. Her deeply creased face was sad, he later told me, but somehow serene. There was a look of relief there—or something like it.

"I'm so sorry," David said.

The woman looked up at the crown again.

"She died almost two years ago now. This was her tree," she said.

David couldn't bring himself to ask for details. Something about her gaze told him not to. He inferred the daughter had grown up here, had loved this tree, had played in its shade, and was now gone, passing on ahead of her mother and the tree itself.

"I'm so sorry," David said again.

"Don't be," the woman said. "It's nobody's fault."

She paused. "I just thought she was safe up there, in the branches. I thought she was safe. I didn't know she was sick again until you came today. I'm so glad you came."

David nodded. He didn't know what to say. "It's so beautiful, this tree," he finally said.

The woman nodded back, silently. David finished removing the last vine a few minutes later and said his goodbye and let himself out through the backyard gate. The woman didn't move. She waved goodbye and kept standing there, quietly, next to that old tree that had just been rescued, her daughter somewhere in the leaves above.

"THERE!" I SAID. "There! Do you see them?"

I tugged at the elbow of my neighbor Lorig Charkoudian,

raising my voice above the gentle sound of wind and the hum of the boat engine. I pointed to the east.

In the distance, two tiny windmills were just now poking their heads above the horizon of the Atlantic Ocean, above the curvature of the earth, under a brilliant September sun. Except they weren't tiny windmills. An hour and a half from now, Lorig and I would be floating right beneath those same windmills, taking in the soft *whoosh* and the brief shadow of each massive arm passing over our heads in the sparkling Atlantic water. Schools of cigar minnows and juvenile amberjacks, meanwhile, would be rising and diving all around us, happily drawn to the "reef" of these miracle windmill machines, everything perfect below a breathless blue sky.

I'll just say it up front: There are only a few moments in this life—only a few—when your body and soul are flooded with the deepest feelings of bliss and well-being. This sunny morning in mid-September, two-dozen-plus miles off the coast of Virginia Beach, Virginia, visiting gentle-giant windmills in the open Atlantic, was one of those moments.

"I see them!" Lorig said now. She gestured with her own hand toward the still-distant wind turbines as the boat chugged farther out to sea and the last of the trailing gulls turned and headed back to shore. Lorig was wearing dark sunglasses and faded blue jeans. On her left hand was a bracelet of tiny birds resembling, I thought, doves of peace flying single file around her wrist. It wasn't a real bracelet. It was a tattoo, in multicolor ink—a kind of permanent reminder of her goal to make the world a better place.

Lorig Charkoudian, fifty, is my representative in the Maryland House of Delegates in the capital, Annapolis. She serves the voters of Takoma Park and east Silver Spring, residing herself in a house just past the Takoma food coop on Ethan Allen Avenue.

She was also, hands down, one of the best clean-energy advocates I had ever met.

But for Lorig to be on this boat, on this particular morning in the mid-Atlantic, was something of a miracle. She was the grandchild of Armenian refugees who had narrowly escaped the 1915 genocide waged by Turkey. She had survived a different kind of trauma in her own life, one of domestic violence, of stalking and assault and bruises, of being thrown to the floor and up against a wall—over a yearlong period by a bipolar common-law husband. One night, after police arrived and photographed her bruises, she took her kids and hurriedly fled to a secretly rented house, ending the relationship forever. And from all of that, somehow, she had risen to become a noted conflict resolution specialist and mediator in her state. She ran a nonprofit, Community Mediation Maryland, in addition to being a legislator.

Now, more than anything, Lorig wanted to help mediate and resolve the ultimate threat of violence: climate change. She had witnessed extreme weather on her own property for many years. A series of rain events sent water into her house, culminating in a basement flood in 2017 that cost her thousands of dollars. Like many people, she installed a new sump pump and lots of indoor-outdoor rugs. But *unlike* most people, she began asking herself what she, personally, could do to change the entire global energy system so that no one, anywhere, would have to face worsening floods and other forms of violent weather. That's how Lorig thinks.

In Takoma Park, her constituents regularly contacted her about a wide range of issues. But since 2018, when she was first elected, the calls about rain and inundated homes had dramatically increased. She got calls about the disappearing trees, too, of course. Lorig herself had lost several trees over the years, including a fifty-year-old oak knocked over, roots and all, by wind after heavy rains virtually liquefied the soil. She and her now-teenage children have set a goal of replanting two new trees every year on their property. They also have a geothermal heat pump for winter warmth and summer cooling at their home.

But windmills—windmills were her special obsession. She thought of them like trees in many ways: sturdy trunks topped with a crown of outstretched limbs that do the work of capturing energy for the organism. There was also that awesome feeling of being in the presence of a modern wind turbine. It was like standing before a giant old tree, just as inspiring, the quiet grace and massive power hushing one's soul.

But before today, before this boat ride, Lorig had only seen smaller windmills built on land, those in the one- to two-megawatt range. She had never been out to view the true giants, the *offshore* types, six megawatts and up. These were the sequoias of the sea, windmills that will soon provide a forest of clean energy for much of the East Coast.

That this was her first offshore trip was doubly ironic since Lorig had done as much as anyone to make offshore wind a reality in the mid-Atlantic. She had passed laws and promoted polices that were accelerating the wind expansion process across the region.

And now, here we were and there they were: the first two ocean-based windmills Lorig had ever seen, getting closer on that September morning. When we were about thirty minutes away, the turbines began to take full shape, twice as tall as the Statue of Liberty, the canopy of blades spinning slowly, elegantly. Just then, a few bottlenose dolphins appeared off the starboard bow as if escorting this seventy-foot charter boat to its destination, to these towers up ahead that might help save the ocean itself one day.

Ironically, our hosts that morning, the ones who had chartered this boat and would feed us a lunch of sandwiches soon, were officials from Dominion Energy, Virginia's electric utility. Known widely as the "ExxonMobil of utilities," Dominion had historically been rabidly pro-coal and pro-gas as a way to generate electricity. It was Dominion, after all, who just six years earlier had wanted to clear-cut the oldest living forest in Virginia—hundreds of three- and four-hundred-year-old trees in Bath County—to make way for a gas pipeline that activist groups like CCAN ultimately stopped. Now, finally, the company was making the giant pivot to clean energy. The two turbines up in the distance were a first-phase step toward a full offshore wind build-out of 176 turbines that would begin right here in early 2024. Ultimately, this wind farm would power nearly 20 percent of Virginia's households.

"Swords into plowshares," Lorig said. "Swords into plowshares." She whispered the phrase several times to me as Dominion technicians, there on the deck, began explaining the nuances of the wind technology to our crowd of invited guests. That Old Testament quote, a plea for peace over violence, for tending to the earth over wounding the flesh, seemed spot-on when a company as notorious as Dominion makes a $9.8 billion investment in hundreds of graceful, spinning wind machines.

Swords into plowshares. Lorig, as a politician, had a way of getting people to put down their swords. It was her superpower, bucking most trends in contemporary American politics. It would be hard to list here all the clean-energy bills she had helped pass at the state level—or supported at the regional and national levels. Her bills in Maryland had names like the Community Choice Energy Act (solar), the Clean Energy Jobs Act (wind and solar), and the POWER Act (offshore wind). With each bill, the mediator in her, the survivor in her, the peace activist in her, would come out. She would get historically pro–fossil fuel labor unions to drop their fears and accept that clean energy really would bring good-paying and secure jobs. She would get environmental groups to stop viewing those same unions as obstacles and instead see them as vital allies. She would get health and justice groups to see that clean energy would lower people's bills and protect the lungs of the elderly and of children.

Above all, she kept her eye on the big prize, that dream of workhorse wind farms in the ocean. Major projects were coming soon to coastal Maryland and New York and Massachusetts. But how do we broaden the wind vision to still more states? How do we use new federal climate dollars and engineering breakthroughs to pick up the pace? What transmission upgrades, for example, were needed to get all those ocean-based electrons farther and farther inland to the suburbs of DC and downtown Baltimore and the rest of the East Coast?

There was, in other words, much work left to do. But today, none of that mattered. Today, on that boat, it was time to see, finally and up close, what Lorig had been fighting for all these years.

WATER STRETCHED AWAY in every direction, no land in sight, when the boat captain idled the engine and we floated the last few feet to a spot directly below the first set of giant wind blades. I looked up. Everyone looked up. The 250-foot-long blades were a brilliant white—white like a Greek town on a cliff is white, someone said—against an equally pure and blue sky.

Then came the sound—a soft swishing sound as each blade approached, getting closer, getting louder, until it passed directly over our heads with the airy-soft sound of a wave washing over a beach. Then the gentle *swish* again, getting fainter, moving away before the next blade came down.

I started to count. One, two, three . . .

In fourteen seconds, that first blade returned, finishing one rotation. In the process, it had created enough electricity to power a typical Virginia home for twelve hours. One rotation.

There on the boat in the bright sun, wearing baseball caps and sunglasses, people high-fived one another and applauded as we reached that first wind tower. One couple in this crowd of Dominion employees and invited advocates and philanthropists hugged and kissed at the bow.

Lorig just kept saying how grateful she was to finally be here—to finally see and *hear* these windmills, and smell the saltwater wind that made them go. For every fallen tree in Takoma Park, for every sodden home and every high school kid in bed with Lyme disease, for every young couple contemplating whether to have children—here was something to hold on to.

For Lorig, the journey to this spot was much longer than for most of us on the boat that day, longer certainly than the 220 miles she had driven from her home in her dinged-up Toyota Prius. It was a journey spanning nearly forty years. In 1985, at age twelve, she wrote a novel about a human society powered entirely, worldwide, by solar panels and wind power and batteries. It was, at the time, the stuff of science fiction. But it was crystal clear in her own mind, a vision she held even before Dr. James Hansen of NASA, in 1988, famously went before Congress and shocked the world by declaring climate change had irrefutably arrived.

Now Lorig, in 2023, was standing right there, staring at the building blocks of her dream world. The wind structures were just as poetic and beautiful as she had imagined—but based in science reality now, not fiction.

A Dominion guide told us the two windmills at this pilot site were 607 feet tall when measured from the base to the top (when a blade is facing straight up). The turbines were rated at 6 megawatts each. But when the full wind farm starts construction here in the spring of 2024, the windmills will be 830 feet tall and 14.7 megawatts in size, each capable of powering a Virginia home for a *full* day with just one rotation of the blades. Together, when complete, this wind farm of 176 turbines would reduce US carbon emissions equal to taking a million cars off the road.

Yet as impressive as all this was, Lorig and I spent nearly as much time looking *down* that day as we did up. If these windmills were the sequoias of the sea, the life below the surface, down among the roots, completed the metaphor with a flourish. Giant schools of silvery cigar minnows were boiling the water all around the trunk-like towers, being chased by equally big schools of Atlantic amberjack. Those in turn were being pursued by long and shadowy barracuda deeper in the water. We could see the barracuda. I counted a half dozen. This food web, stimulated by the hard cover of each windmill's underwater foundation and its plankton-rich reef, attracted larger species all the way up to the hammerheads and sandbar sharks and giant manta rays known to prowl these waters.

"Imagine when this wind farm is fully built out," one Dominion employee told me. "It will be one great and long and wide reef for marine life."

Recreational fishermen here were ecstatic about the windmills, for obvious reasons. Commercial fishermen grumbled about the restrictions on fishing near the turbines but had generally accepted a future of cohabitation.

Over the past twenty years, a host of European countries have constructed dozens of ocean-based wind farms like the one coming here. Scientists, meanwhile, have gathered encouraging data on their impact on wildlife. When properly sited—and with noise-reduction methods used during construction—these farms have posed no meaningful threat to bird or marine species, including to migratory whales. Constant vigilance is needed in the future, but it appears these living creatures—of the air and of the sea—are pretty smart. They mostly fly or swim over and around the turbines as they would an island or a large boat. And many of them profit from the miraculous reef.

But it still felt surreal to me that Dominion Energy was actually doing all this here in America, moving from the violence of fossil fuels to the world of wind energy. Swords into plowshares. It was a giant step forward, to be sure. But it left me pondering, inescapably, the bigger picture. Even during this perfect moment, on this celebratory ship, in the full sunshine of maximum content—even here, I couldn't forget the central fact: it wasn't enough.

As 2023 wound down, destined to be the hottest year ever measured—and as my yearlong quest for new climate answers wound down, too, with autumn's approach—I was amazed all over again at how long we had waited to even *start* implementing big solutions like offshore wind.

Lorig, my hero, was going to keep pushing for more and more wind and solar power for the rest of her career, till we reached a carbon-free economy in the decades ahead, by god. But for this planet to actually survive, stretching from the microcosm of Takoma Park to the vastness of this ocean ecosystem, the list of add-on solutions really was daunting. People like Ning Zeng truly did have to step it up—fast—in their multifaceted quest for carbon sequestration solutions. I grew increasingly worried every day about the pace of that work. And people like Jamie Raskin, in Congress, had no room for error in their mission to preserve a functioning democracy in America capable of processing the facts of climate change. No room for error. And most important, perhaps, people like NASA's Tianle Yuan were going to have to figure out how to give some of this brilliant sunlight back to the cosmos—the sunlight flooding our boat that day and shining down on the Atlantic Ocean and beaming down upon this whole earth—some of it likely had to be reflected back to outer space. No lack of worry about that either.

But if anything was working in our favor, I realized, it was this: the sheer will to live. Beyond the riot of fish on display that morning, our boat captain told us about the healthy populations of loggerhead and leatherback turtles that inhabited these waters, rising often and feasting on jellyfish. And then there were the birds we saw that day: sanderlings flying in dense flocks close to shore, a red-necked phalarope flying directly overhead, even a dark-eyed junco songbird that must have gotten blown offshore and landed on the boat and caught a ride all the way back to shore with us.

After all we've done to pollute and abuse our oceans and continents, then adding climate change on top of all of it, the great majority of the world's life-forms were still stubbornly hanging on, at least for now. I saw it in miniature every day on my Willow Avenue block and surrounding streets as brave saplings of oaks and poplars and maples rose up to fill new tears in the neighborhood canopy, the trees doing all they could to weather the heat and floods and high winds that kept bringing down their elders.

The climate leaders I had trailed for much of this year had it, too, this remarkable will to live and survive and rise above the hurts of this world. Ning and his family had survived the poverty and torture of the Cultural Revolution. Jamie Raskin had endured cancer treatment and the loss of his son and the attack on January 6, when literal swords and clubs came out.

And then there was Lorig, a single mom and survivor of domestic abuse with all its residual burdens. And still, seven days a week, she got up and placed the *public* interest ahead of nearly everything else in her life. Her grandparents, meanwhile, had taught her that survival, even of genocidal attack, against staggering odds, was possible. Lorig now saw a different form of extermination looming over all of us—ecocide—and she would fight with all she had, for as long as it took, to stop that outcome from canceling out the world of her once-dreamy childhood novel, a world so heartbreakingly close to reality now if we could just hang on.

The boat had just turned around that morning, and we were heading back to shore, the windmills getting smaller behind us, when I asked Lorig the question I'd put to so many others: "Are you hopeful?"

There was a pause. She looked down at the water. "I don't

know if I think about it that way," she said. "I just think: What can I do today? What can I do better tomorrow? What did I learn yesterday? And I think, *No matter what, we can't give up.*"

The distant shore of Virginia Beach, a smudge on the horizon, was slowly coming back into focus to the west. The bottlenose dolphins were returning, too, many more than on the trip out. There were a dozen of them off to the port side, including some of the largest adult dolphins I had ever seen, weighing close to five hundred pounds, the captain told us.

And there were babies. Several adults were paired with juveniles clearly born that summer, their tiny dorsal fins rising, curling, and then falling below the sea.

10

Grave Site

October

Ning Zeng wasn't going to be a criminal after all. He wasn't going to face legal jeopardy just for burying trees in the fight against climate change. By October, he had finally figured out a way to entomb one hundred metric tons of deceased trees in Maryland, legally and immediately, with the hope of thousands more tons to follow soon.

So Ning was feeling pretty optimistic as autumn arrived in full in the DC region. October pumpkins and mums and skeletons now covered many suburban lawns and urban porches in the area. But along many streets, trees continued to die in real life and turn into real skeletons. The unusually dry weather didn't help. Rainfall levels had been below normal for much of the year—and high temperatures in September and October set records for many dates. Abnormal weather had become so *normal* here that kids preparing to trick-or-treat had never known anything but strange heat and scary deluges and sudden droughts—and the resultant sound of chain saws taking down trees somewhere in the distance.

Ning's job now was to prepare a grave site for at least some of those deceased trees. The farm in Cecil County was ready to receive the first bodies of oaks and maples and poplars, pending a few final arrangements. The trees were coming from Camp Small, the transfer station in Baltimore where I had visited with Ning back in February. City leaders in Baltimore had maintained their pledge not to send deceased trees from municipal property to the city's landfills or trash incinerator. So those trunks and limbs, brought down by old age or bigger storms or climate-related disease, continued to pile up at Camp Small's multi-acre transfer lot. The bodies were still stacked so high they formed small mountain ridges with valleys below where logging vehicles loaded and rearranged trees along dirt roads.

And now there was a new problem: fire. One consequence of the ever-growing volume of trees was that the woody mulch pile on the northern side of the yard grew taller and taller. To try to reduce the pile's size and the ever-present risk of unintended combustion, the yard managers were *paying* landscapers and tree-care companies to haul away the mulch for various commercial uses. But the glut of wood mulch was so great throughout the region (from rising mortality and ongoing construction of buildings) that even the seventy-dollars-per-ton premium paid for mulch removal was only slowly denting the Camp Small stockpile problem. Demand for the yard's wood just wasn't there.

Then, in September, it happened. The giant pile of mulch—a cooking mountain of hot, decomposing biomass—caught on fire. It was spontaneous combustion, the kind that can happen with big compost piles. Smoke began rising from the summit in mid-September, coming up from the smoldering wood below. Eventually, the whole yard was shut down out of fear the fire could spread disastrously to the mountains of stacked logs all around.

The Baltimore City Fire Department was called in to try to put out the stubborn fire, but the fire kept going for days until finally, in early October, it petered out. Meanwhile, the supply of newly dead trees never stopped coming from Baltimore's streets and parks and other public lands. For much of September, because of the fire, that wood was diverted to a back corner of the city's beloved Druid Hill Park two miles away, where a new, long ridge of trees stacked twelve feet high began to grow.

So Camp Small's headaches were worsening. Its latest strategy of paying landscapers a premium to haul away tree debris had, well, gone up in smoke before the wood could be fully removed. The beleaguered yard needed Ning, more than ever, to please just take the damn trees as fast as possible.

And god knows he was trying. Having been stymied by red tape and slow-moving bureaucrats, his goal now was to get legislation passed by the Maryland General Assembly in early 2024 to make what he was doing legal across the state. In the meantime, he had discovered that local laws in Cecil County permitted him, with a few wrinkles, to bury a modest quantity of trees with no additional state permits required. So his goal was to put one hundred tons in the ground in the fall as a demonstration project and then thousands more tons underground in the spring.

The trucks Ning needed to transport the trees were not technically hearses, but they would serve the function: to transport the deceased plants to the first-ever cemetery in the world "at commercial scale" that used buried trees to sequester carbon. The logistics of transporting just one hundred tons of sawed-up trees forty-nine miles to the farm were challenging. It would take three large transport trucks two days, making ten total hauls. Ning would also have to buy a $4,000 mobile truck scale to make sure his cargo per truck did not exceed state highway weight limits.

Despite the laborious transport process, Ning's calculations showed that 95 percent of the trees' net carbon content would wind up in the ground even after subtracting for CO_2 released in transport and burial.

Ning was still on sabbatical that fall and did most of the logistical planning for the burial project from his home office and laboratory in east Silver Spring, bordering Takoma Park. While his neighbors were putting spooky tombstones and coffins in their yards in October, he was designing his own cemetery and thinking about the special ceremony he would have when the first trees went into the ground. He had a Chinese poem in mind about grief and the eternal peace of the afterworld. He would read the poem and then there would be a moment of silence, he said.

Meanwhile, hands down, the eeriest news of all that fall came from climate scientists tracking average monthly temperature records around the globe. The Japanese Meteorological Agency was the first to release data showing that September 2023 was a staggering 0.5 degrees Celsius (0.9 Fahrenheit) warmer worldwide than the previous warmest

September on record (2020). That number—0.5 degrees—was also *by far* the greatest temperature rise for any one month above a previous monthly high. When you look at the historical temperature graphs, you see September 2023 just burst upward out of the pack like a cannonball shot into the sky. It looks, quite frankly, like a temperature explosion.

News of this extraordinary September spike left me shocked all over again by the sheer speed of the planetary warming. All the more reason to bury as much carbon as possible—as fast as possible—under the earth's surface. So when Ning extended a special invitation to me that October, I readily accepted. He asked me to join him and some of his students in traveling to the Kilby family farm, in Cecil County, Maryland, to survey the land and pick out a grave site.

So MANY ACORNS fell that October in my backyard, tumbling down from our gallant pin oak, that the nuts overwhelmed all the animals living near the tree's big crown. The squirrels, deer, white-footed mice, rabbits, raccoons, voles, crows, blue jays—acorn eaters, all—couldn't keep up. The manna was too much, falling daily. Even the hoarders among them, critters specializing in storing excess acorns, were outmatched.

And that was precisely the tree's strategy. It was a so-called mast year in 2023 for many of the oak species in this region. That's when the trees periodically heave forth many, many more acorns than in a normal year, and thus the survival rate of baby trees, from acorns to seedlings, dramatically increases from that year.

Mast. Some arborists insist it's derived from the word *masticate*. A big meal was happening for all the critters in my backyard. The squirrels, standing upright and clutching acorns with front feet, ate and ate and ate. Then they started digging—everywhere—hiding the overflow acorns in holes all across my property. The brains of these eastern gray squirrels actually *grow* as much as 15 percent larger in the autumn, allowing them to memorize the location of hundreds of underground caches for winter consumption.

The blue jays were digging, too, their relationship to oaks even older than the squirrels'. Oaks and blue jays trace their evolution to the same small region in what is today Southeast Asia. For sixty million years, they have evolved and spread symbiotically across much of the planet. The North American jays in my backyard have an esophagus perfectly shaped to hold as many as five acorns at one time. And what they don't eat now, they bury. Each jay is capable of interring more than three thousand acorns in a single season.

So the digging that fall was prolific all along Willow Avenue, at least below or near the mature oaks still standing on the block. I watched the ground in back of my house as the leaf cover went flying and the scratching and digging began. And as more jays and squirrels went to work, I realized something: they were creating the opposite of a graveyard. It was a cemetery of new existence. Whatever feelings of sadness and resignation surrounded Ning's novel mission to bury deceased trees—those feelings were nonexistent on these golden afternoons of autumn as I watched a natural army of gravediggers spread out, creating a happy burial field of newly planted oak nuts across my yard.

Meanwhile, the baby oak tree already hovering in that yard, the one I managed to save in July after nearly killing it while cutting grass, was now five inches tall and still growing and looking better every day. It was protected from deer by the cylinder of chicken wire fencing. I decided I would likewise protect and defend as many of the new mast-year oaks as I could, watching out for those seedlings that rose up from the cemetery next spring, acorns to tiny trees.

How cruel, though, that deer were one of the biggest threats to these oak seedlings, chomping on the tender leaves as they emerged in spring and summer. Those deer actually needed the trees to survive. Acorns provide up to 25 percent of their diet in the fall, the high-fat content prepping the deer's bodies for the sparse food supply of winter. But like the weather, the deer population remained completely out of control here, with lethal consequences for oaks.

People died, too. October brought the start of the annual rutting season for deer in Takoma Park, with the absurd sight of antlered bucks chasing does across front lawns and into streets and across nearby highways. My friend George Leventhal somehow survived when, several years ago, a sizable buck came crashing through his driver-side windshield one November night—on the *Beltway*. What will it take, I wondered, before we humanely cull these white-tailed killers?

And during that autumn of 2023, my infant oak out back, all fenced in, faced yet another terrible attacker: drought. What had been a drier-than-normal year prior to October turned into a full-on mini drought by late October. The spigot just shut off. By Halloween, just 0.66 inches of rain had fallen in the DC region for the month, the tenth-driest October ever recorded here. Distressed, the mother pin oak in my backyard began dropping small branches again. Farmers across the region reported creeks and rivers too low to use for irrigation.

Quietly, with no rain falling and no other option, the parched oaks of my neighborhood and the surrounding region began relying on stored carbohydrates in their trunks to maintain basic tree functions. It was the same stressful chemical process that many trees employed in 2019 when the phytophthora mold destroyed their roots from *too much* rain. That chemical process of cannibalizing internal nutrients produced, unavoidably, the volatile organic compound ethanol, vented into the air from the trees and serving as the dreaded catnip to ambrosia beetles. Many arborists in the region were now concerned that, absent a return of rain soon, another beetle outbreak could occur. Too much rain or not enough, the beetles didn't care. They were ready to pounce during whatever weather extreme tumbled off the seesaw of climate change.

Long term, the overall trend in this region was clear: warmer weather would bring greater annual rainfall and bigger individual precipitation events. Droughts, though damaging and periodically deep, would come and go. But the climate models showed wetter areas of the world, like the US East Coast, will continue to get wetter and wetter as the warming air above absorbs, holds, and then releases more moisture supplied by the world's faster-evaporating oceans. Indeed, rain now seemed just around the corner for the DC region, with forecasters pointing to the El Niño weather phenomenon then forming in the Pacific Ocean. That El Niño would likely bring rainier-than-normal conditions starting in late autumn in the mid-Atlantic, scientists said, continuing through the winter and maintaining our overall pattern of sogginess.

But for now, it was dry. Really dry. Autumn leaves crackled and crunched underfoot with deafening loudness, no moisture to dampen their voice. Grass and shrubs died. Birds suffered. And almost every day, a blue sky hovered overhead, perfectly cloudless, unbearable in its carefree autumn beauty as the land groaned below.

My friend Barbara Briggs owned that special gadget she used to test for gas leaks underground that can harm tree roots. But no one I knew had the kind of highly sensitive listening device—the one for detecting vibrations coming from dangerously dry and distressed trees—to determine whether all those tree trunks in Takoma Park were, in fact, screaming that October from the drought. Even if I had such a device, I wouldn't use it. I didn't want to know.

To relieve the suffering of my own pin oak, I began watering it as best I could in late October. I followed an arborist's instructions to use a drip hose to soak the ground all along the outer edge of the canopy.

As for the baby oak, I continued my de facto role as midwife, applying water gently every few days from a two-gallon watering can while obsessively checking the chicken wire cylinder protecting it from deer.

Not that the deer or squirrels needed anything else to eat that October. On and on, the acorns came pouring down that month, unlike the rain, falling in buckets. By late October of that mast year, Beth and I were ducking and sprinting during trips to the backyard compost bin or on errands to the toolshed. The acorns bounced off our shoulders. They pinged and dinged off the metal gutters and wooden porch steps. An acorn landed in my coffee one morning. Acorns covered the ground.

It seemed impossible that one tree, this seventy-year-old pin oak, in the middle of a dry spell, could do all this. But its canopy was now the second largest on the block. And maybe it sensed, this tree, that the long-term odds were stacked against it—that maybe a mast year to end all mast years was its best chance to sow its genes into the soil. On cue, the loyal jays and squirrels were there to assist, digging their nurturing holes, the burial plots spreading in all directions. And a pair of human hands, in the spring, would be there, too, coaxing and watering and applauding the new trees.

THE FALL COLORS were just beginning to pop that mid-October day as we drove to the Kilby farm, ready to select a final resting place for Ning's trees. The warm and dry weather had delayed the foliage shift, but gentle eruptions of red and orange and gold leaves were now happening everywhere, making the country roads of Cecil County, north of Baltimore, quite lovely. We passed old farmhouses and horse paddocks edged by the shifting autumn hues. We passed meadows of goldenrod, too, blooming brightly despite the dryness, and cornfields harvested early, the stubble left yellowing on the ground.

Ning and I were traveling that day in the back seat of a Honda Odyssey minivan. It was driven by his former student Jazmine Escobar. She had volunteered to transport us and two other students, Dillon Capalongo and Yasmine Tajeddin, to the farm. But Ning wasn't looking out the window at the scenery much that afternoon. He was too busy texting the engineer already waiting for us at the farm and the farmer himself, Bill Kilby. They were wondering where the heck we were.

Turns out, our first stop of the day had taken longer than expected. We'd spent the morning collecting data at Camp Small, the overwhelmed tree yard in Baltimore. While Dillon flew a drone camera overhead, documenting the insane maze of trees, Jazmine and Yasmin had helped Ning measure the moisture content of a sampling of trees scheduled for burial. Knowing the moisture level helped Ning more accurately quantify the carbon content that would soon be locked away, by design, for a thousand years.

Now, as we got closer to the farm, sitting in the back seat of the minivan, I asked Ning how he felt. His dream was about to come true, at least the first step.

"I don't know," he said. "I'm too busy to feel excited right now."

He'd been working eighty-hour weeks, missing family events, losing sleep. He'd been negotiating with trucking companies and excavation companies, doing stuff he never thought he'd have to do as a scientist. One night, driving back late from Cecil County, he got into a shouting match on the phone with one of his consultants until the car behind him on the highway began honking for him to speed up or pull over, offering a finger for punctuation.

Finally, that October day, we turned off Firetower Road and entered the property of the Kilby dairy farm, six hundred acres of feed corn wrapped around an indoor milking facility and feeding pens. Ning directed Jazmine to turn left into one of the cornfields—now harvested and replanted in winter rye. We followed loose tracks through the field and down to a stretch of bottomland along a creek called Basin Run. Moments later, we filed out of the minivan to the sound of cows in the distance and the faint gurgle of the nearby creek.

Scott Petrey, the engineer Ning hired to help plot the precise location of the burial spots, was already there, standing beside a grove of maple trees and sycamores and black walnuts lining the creek. He greeted us with a survey pole in one hand. It was topped with some sort of GPS device ready to talk to satellites and record the chosen coordinates for the graves.

Once out of the van, Dillon launched his drone again, its tiny propellers buzzing overhead while filming the lovely, sloping fields of the Kilby farm. Just then, an old white pickup truck came bouncing down the tracks to our rendezvous spot. It was Mr. Kilby, age seventy-eight. Everyone called him "Mr. Kilby." He stopped, got out of the vehicle, and walked toward us, passing the truck's front license plate, which read, "Our farms. Our future." He was dressed in boots, jeans, and sunglasses. He had a thick white beard.

Ning was the first to greet him. They shook hands. The men had met several times before here, but you would have thought this was a first greeting the way Ning lingered. He had a way of shaking hands and not letting go sometimes, of clasping firmly and holding on just a while longer as if saying, "I see you, I really see you, and I want to keep you close another moment while my full respect for you fills the space around us." That's what happened with him and Bill Kilby that afternoon.

I stepped forward and shook Mr. Kilby's hand, too. I asked what made him want to be part of this pioneering project, bringing the commotion of mass tree burial to his peaceful fields.

"Son," he said, skewering the earnest tone of my question, "you must not understand how boring dairy farming is."

Everyone laughed. In truth, Bill Kilby was quite an outlier among US farmers, constantly innovating on his land to create greater sustainability and reduce his operation's carbon footprint. The manure from his eight hundred cows was composted in a biodigester on the property that captured the methane and converted it to electricity to power the whole farm. He practiced no-till farming on his fields, locking more carbon into the soil. And for decades, he had been planting new trees all along Basin Run, creating a forest buffer along the creek banks that shaded the water and lowered its temperature enough that, twenty years ago, the native brown trout started reproducing naturally again.

And now he wanted to store five thousand tons of carbon on his fields in the form of buried trees. He was donating two acres of land to Ning for the purpose.

Ning and Mr. Kilby began walking, side by side, along the rye field north of the tree line. The rest of us trailed close behind. Mr. Kilby exhibited the energy and stride of a man much younger. Only his voice, deep and gravelly, seasoned by years of outdoor work among loud tractors and talkative cows, gave away his age.

On they walked, Ning and Mr. Kilby, quite the pair: the rural farmer from a politically conservative corner of the state, and the Chinese American climate scientist who spent most of his days in front of laptops and chalkboards. But the chemistry between them was real and obvious—in part, I think, because Ning, too, had farming in his blood, descended from thirteen generations of rice growers.

With input from the engineer and after questions from the students, Ning and Mr. Kilby finally selected a plot sixty feet long and fifteen feet wide to receive the first wave of trees. The plot was carefully chosen to lie outside the root zone of the creek-side forest. Ning explained how the delivery trucks from Camp Small would arrive first, stacking the tree bodies in rows adjacent to the marked burial pit. Then, if all went according to plan, on Monday, November 13, an excavator would dig the grave and lower the trees.

Until today, I hadn't fully understood the details of Ning's excavation and burial plan. There were many possible ways to entomb trees for carbon storage, he told me. One of the best was underground in super-dry desert conditions where there is virtually no moisture to aid decomposition. Even better, he said, was the opposite: burying trees in waterlogged conditions with high-quality clay. Biology struggles in extremes, he said, whether super dry or super wet. After all, some of the best-preserved Viking ships and ancient trees and early humanoid footprints began with accidental submersion into the mud of old river bottoms or in oozing wetlands. Which is why, for this maiden project, he had chosen this bottomland, next to a creek, where the clay content was high quality and the groundwater could be trapped in place. The muddy, clayey, wet, stagnant conditions, with nearly zero oxygen, would shut down virtually all biological activity.

"This," Ning said, "is how you convert biology to geology. Normally, it takes thousands of years. We're going to do it here in a few hours. The carbon in these trees will be locked away underground with a thousand-year stability pretty much instantly."

That certainty and quickness were vital given that climate change, seen right here on this farm, was also certain and quick—in the wrong direction.

Mr. Kilby was walking back to his pickup truck, all of us preparing to leave, when—in his husky voice—he told me what he'd seen over his lifetime here.

"When I was a young man, autumn came to this farm two weeks earlier than it does today," he said. "And spring came two weeks later. Think about that. We get a full month more time to farm now. So initially, I think the warming helped us."

We reached his truck, and he leaned against the cab door for a moment, the waning afternoon light falling all around us.

"But now? Ten years ago, we lost all our ash trees on the

farm. They just all up and died, killed by those ash borer beetles that I'm told are connected to warming. And the rain keeps changing, which should make every farmer nervous as hell. It's dry now, but two years ago, we got two three-hundred-year floods in one year. You take all that together, and I don't see how you don't see us getting really hurt eventually. So Ning better not screw this up. We all know it's just one step—one out of the many we need to take—but if people are farming here a thousand years from now and we've whipped this warming thing, maybe there will be a plaque or something that says part of it started right here on the Kilby farm."

IT HAD ALREADY been a long day as we climbed back into Jazmine's minivan. A chilly wind had kicked up, blowing across the farm. Everyone was tired.

From the back seat, Ning checked the clock on his phone. He seemed pleased with the day's work but also anxious. He looked out the window and spoke his mind. "More land," he said as we pulled away. "I *have* to find more land. More farms."

The excitement of the day was already giving way to Ning's near-constant obsession with burial space: How could he get enough of it, with the right clay, for all the Camp Small wood and for other trees dying across the region? Just recently, he had learned of a new potential source of burial trees: driftwood logs were piling up behind the large Conowingo Dam, just six miles from here on the Susquehanna River that flows down from Pennsylvania. Driftwood accumulates behind all dams, even under normal conditions, but there was growing anecdotal evidence that bigger rain events linked to climate change were washing more and more riverbank trees into eastern rivers in recent years, drowning the trees, their bodies congregating dangerously behind dams like the Conowingo. They needed to be buried, too, Ning said. But to do that, he needed land.

So as we headed home that day, he told us he'd been quietly making multiple trips up here to Cecil County in recent months, acting as a kind of land prospector, hunting for suitable property. He was armed with special geological maps showing the rough location of the desired clay—*smectite* was the soil classification—running just below the surface in large tracts across the county. With the help of a local real estate agent, he had been looking for farms for sale on land overlapping the clay. Most farmers would not, Ning knew, donate their property like the conservation superstar Bill Kilby. He would have to buy the land using money from sales to the international carbon-offset markets.

But none of his prospecting had so far panned out. Either the size or cost of the lots didn't match what he needed. One day, when he thought he'd found a winning plot, he decided to test the soil to confirm it matched his map. Without permission from the absent land owner but with a nod from the real estate agent, he took several rogue soil samples, quickly drilling down with a mechanical coring device. He found clay, all right, but then found a stubborn layer of rocks several feet below the surface. It was a bad site. He had to keep looking.

Which is why Ning made a special announcement that day, there in the van, as we headed south from the Kilby farm toward the interstate. "The day's not over, students," he said. "We have one more farm to visit. I want to make a cold call."

You could almost hear the collective groan in the car. The students turned to him with exhausted faces. "A what?" Dillon asked. "What do you mean 'cold call'?"

Undeterred by his recent failures, Ning had added cold-calling to his tool kit. He was bypassing the real estate agents to speed up the land acquisition process. If a property lined up with his soil map, he would just walk right up to a barn or a farmhouse and knock on the door and explain himself. Unsurprisingly, given the altogether odd nature of the proposition, delivered out of the blue by a college professor standing at the door, this method had yielded no big breakthroughs either. But Ning wasn't giving up.

He was, however, getting tired of traveling all the way up to Cecil County to bury Baltimore trees. What he wanted more than anything now was to find a farm much closer to Baltimore for the Camp Small supply. This would cut down on the obvious cost of transport and the attendant carbon emissions. He had chosen the Kilby farm for his first burial because the land was donated. (He had met Bill Kilby through a mutual contact in state government.)

So now, in the van, Ning told us we were going to the Zahradka Family Farm in Essex, Maryland, a place just east of Baltimore on the Chesapeake Bay and only eighteen miles from Camp Small. His soil maps showed the farm rested atop the right kind of clay. Also, like the Kilby farm, it was a designated *conservation easement*, meaning it would remain farmland forever, never sold for development.

"But have you called them?" Dillon asked, looking back from his front seat. "Do they know we're coming?"

"No," Ning said. "There was no phone number on the website."

"So we're just going to walk right up and explain we'd like to bury trees on their property?" Dillon said.

"Yes," Ning said.

I realized then, for the hundredth time, that Ning's ability to think outside the box wasn't the only thing that made him a potential incubator of mass change. And it wasn't just his enormous courage, investing everything he had in a single idea. It was, again, his simple will to keep going, to never stop, to do everything large and small, even if his steps sometimes made questionable sense or were such long shots as to make you scratch your head. His wife, Annarita, told me later that Ning, in her mind, personified the ancient Chinese proverb about how, with enough time, dripping water will wear down even the mightiest stone.

Ning put it a different way, often talking about the grit he picked up as a child. He sold sweet potatoes on the street at age twelve when his family was very poor in China and needed every penny. And he and his brother often skipped school and walked the two hours to their grandparents' house, catching fish in rice paddies for food, hunting birds with their slingshots, presenting everything to their grandparents when they finally arrived at the tiny hut under the giant red bean tree.

WHEN WE PULLED up to the main buildings of the Zahradka farm in Essex, no one was there. No humans, I mean. Dogs were everywhere. I counted eleven dogs sleeping, running, barking, and lazing about. There was also a menagerie of baby animals behind a fence with a sign: "Pet at your own risk." An adorable calf ambled up to us as we got out of the van. Then baby ducks and a lamb and a little goat.

Twenty minutes earlier, we had left interstate I-95 near Rossville, following roads through sprawling suburban neighborhoods until, getting closer to the Chesapeake Bay, the roads became narrow two-lane affairs running through thick patches of forest and over backwater creeks. Signs to the farm then led us to this dusty gravel parking area surrounded by the animal pen, two barns, several tractors, and a big outdoor vegetable stand teeming with that year's final sweet corn and newly plucked pumpkins. But no people.

Ning stepped out of the van and strode directly to the "Purchase here" sign on a table at the back of the extensive vegetable stand. He didn't pause to check out the acorn squash or bouquets of yellow mums before projecting his voice loudly: "Anyone here? Anyone from the farm here?" He waited a short moment before saying it again.

When no one came, he walked around the purchase table and into the small barn just behind the stand, the double doors open. I followed him as he continued to call, "Anyone from the farm here?"

There were empty bushel baskets and tools and wooden pallets stacked against the barn's metal walls, but no people. Ning kept going, passing through a set of back doors leading directly to a half dozen fenced-in cows out back. The cows stared dully at him as he yelled in several directions across the surrounding fields. "Anyone from the farm here?"

"There's a house over there," I said, pointing past two pickup trucks to a two-story brick structure.

Ning headed that way and got within twenty feet of the door before two dogs emerged to stop him. They weren't really the guard-dog types, wanting mostly to be petted by the sudden stranger. "Hello? Anyone from the farm?" he yelled toward the house, stymied by the happy dogs.

The door swung open, and Libby Zahradka—co-owner of this farm with her husband, George, mother of two teen boys, owner of lots of dogs—said to the pets, "Get away from that man! Get!"

The dogs scattered as Libby walked over and shook Ning's hand, saying in a friendly voice, "What can I do for you?"

Without much chitchat, he launched into his cold call, talking about climate and carbon and trees. I watched as Libby, forty-one, her black hair pulled into a ponytail, was put at ease by Ning's academic demeanor that came across as nerdy, for sure, but also utterly friendly and charming somehow.

When it became clear this was going to take a few minutes, Libby waved for Ning to follow her. "I've gotta feed these cows, but you can keep talking." He did, quickly getting to the point.

"My map shows that you have really good clay underneath your fields. I think this could be an excellent place to bury trees."

"Oh, we know about the clay," she said, now tossing broccoli stalks—left from last week's farmers market in Baltimore—toward the same cows we'd encountered earlier. "We have to add sand to the soil when we plant carrots and anything needing soft soil."

She also said the clay harbored arrowheads and, in some places, huge waste piles of oyster shells left by indigenous people. "Why not add trees, I guess, if it helps the environment?" she said.

Ning's students had joined us by then and were staring at each other: Was this a breakthrough? Was Ning really pulling this off? Libby took the professor's card and said she'd talk to her husband, who was sleeping that afternoon after some tough workdays, and maybe Ning could come back later and provide more details.

"And did you want to buy anything?" Libby asked.

The van was fairly stuffed with pumpkins and a giant bouquet of mums as we headed back to the DC suburbs, the day finally, really over. By all indications, Dr. Zeng had in fact pulled off a promising cold call. Everyone congratulated him as the last light of afternoon passed from the sky on a day that had begun long ago at Camp Small.

Ning announced that no matter what happened at the Zahradka farm, he wasn't giving up on Cecil County. He had found, again, a potential new source of wood nearby: dead trees piling up behind the large hydropower Conowingo Dam on the Susquehanna River, washed there in growing volume by the rain and floods of recent years.

"How many trees wind up behind the dam?" I asked Ning, growing increasingly curious.

"Thousands of tons per year," he said. "A lot. But it's much more when there's a bad storm year."

In this wetter region, driven by warmth, it was inevitable: rivers were jumping their banks more frequently now, rising to record heights, reaching deeper into riverbank forests. Urban and suburban trees might succumb to the phytophthora mold after sustained drenchings, their roots destroyed. But trees living along countless rivers and tributaries—literally thousands of miles of waterways in upstream Pennsylvania and Maryland alone—risk having their roots just ripped out of the ground entirely by roaring water, their bodies tossed into the torrent.

Constellation Energy, the company that owns the Conowingo Dam, the largest dam in Maryland, had in recent years increased its budget for capturing the accumulating logs—as much as five acres' worth floating behind the dam on some days. The company barged the trunks and limbs upstream, where they were ground into wood chips or turned into compost for gardeners, the wood rapidly converting to CO_2 as it decomposes. Ning was ready to negotiate with the company to allow burial of the trunks instead on Cecil County farms—if he could find more sites.

Of course, during *really* bad storms, the trees don't pile up behind the dam at all. To prevent catastrophic dam failure, the flood gates are opened, and the roaring water is allowed to pour out. The trees caught in the water, many still technically alive, barrel over the tops of flood gates, falling eighty feet to the racing river on the other side, headed for the open Chesapeake Bay.

In 2018, the year all the states in the mid-Atlantic set records for high annual rainfall, triggering underground phytophthora growth and subsequent mass mortality of oaks—that same year the Conowingo Dam was opened up twice, in July and November. A staggering area of the Chesapeake Bay then filled up with tree debris, clogging marinas and covering beaches from Havre de Grace to Baltimore to Annapolis. The gnarled and floating trunks and branches came from many of the bay's tributaries, but especially the Susquehanna, its biggest, running through Pennsylvania into Maryland. It was one of the biggest debris events of its kind in many years on the bay. It took months to remove hundreds of thousands of tons of trees from the water.

"The next time that happens," Ning said, "we have to capture as many of those trees as we can."

It wasn't *if* that ever happened again.

Ning said, "The next time."

The current dry period in this region would soon end; the wetter El Niño weather pattern was approaching—and who knew when the next record rainfall year would arrive here given 2023's delivery of such soaring new high temperatures worldwide?

So on the drive home that late afternoon, despite the day's

accomplishments, I began to feel a familiar pang of sadness in my gut. And I felt bad about feeling bad. It had been a good day, a hopeful day, spent on two happy farms—so why couldn't I be happy? The answer, I knew, was that happiness doesn't linger long when you think too hard about any one thread of the world's unraveling ecological fabric. I couldn't stop thinking about a future terrible storm in this region and all those trees washing down the Susquehanna River, falling over the dam, their bodies tossed violently into the bay. Some of those trees might float as far as the Zahradka farm in Baltimore County and then to bay shores everywhere and to people everywhere. And maybe, I thought, in desperation and in decency, having heard Ning's voice somewhere, somehow, people will drag the trees out of the stormy water, pulling them up to higher ground, quickly burying them, almost randomly, everywhere, side by side with the seeds, with the scattered, vulnerable acorns, in the cold, deep clay of this upside-down landscape.

THERE WAS NO one in the world I respected more than Ning Zeng. He was the smartest scientist I had ever met. He brought the highest possible integrity—personal and professional—to a vision that allowed him to peer deeply into the future.

So it was hard for me, very hard, to confront him that Sunday afternoon, just two weeks before the big burial day, about something I saw as a serious flaw in his work.

The task of leaving my house and heading for Ning's that afternoon was made harder by the sheer beauty and allure of the autumn weather. Few days are prettier in my neighborhood than during that final stretch of October when the willow oaks on Willow Avenue drop all their willow leaves. I just wanted to stay home and watch it: the burnt-orange leaves floating beneath a low sun, through light that seemed passed through honey, until reaching yards and sidewalks below. All eight willow oaks were dropping their goods that afternoon, the leaves like small cigars, oblong beauty, floating to the ground. A single mature oak can shed up to seven hundred thousand leaves in the autumn. So it was snowing—like, seriously snowing—up and down the street that Sunday afternoon as I departed, the glowing orange leaves spinning and corkscrewing through the honeyed light, covering everything.

It was snowing at Ning's house, too, as we took seats on his back porch beside a band of red oaks and maples and poplars. The trees and their falling leaves separated Ning's backyard from a pair of baseball diamonds in a lovely park next door, Nolte Local Park. I coached my son's youth baseball teams there for years. Ning's wife, Annarita, was now intent on saving these nearby trees, with clippers and pruning saws, from a stubborn outbreak of invasive vines. One tree she couldn't save was the white oak that died in 2019 from too much rain and the subsequent beetles. The stump sat just below Ning and Annarita's porch, always visible, like an urn of ashes somehow on a family's mantel from a past war, where almost every clan in the neighborhood had lost at least one family member to the battle.

The confrontation that afternoon with Ning began with a piece of paper covered with serpentine lines. I placed the paper on the porch table, between his chair and mine, so we could both see it.

"Have you looked at these temperature numbers closely?" I asked Ning.

He stared at the paper. It was a graph composed of sixty-five squiggly, horizontal lines, a blur of lines, representing the average annual temperature—month by month—for each year on Earth since 1958, as recorded by the JRA-55 measuring system, one of the best in the world. Over the last thirty years, those lines could be seen trending significantly upward. Over the last ten years, even more so. And in July 2023, August 2023, and September 2023, it was a horror show. The lines jumped up between 0.25 and 0.5 degrees Celsius from previous monthly records. In the earth's past, it could take many decades or centuries to see temperature shifts that great. Now it was happening in a matter of months.

"People just aren't talking about this enough," I said to Ning. "These temperature numbers should be on cable news every night and on front pages. It's completely insane, much more than many models have predicted."

There was a pause as we both looked at the paper.

"I know," Ning said.

Ning and I had spent a lot of time together since we first connected that cold day last February back at my house, touring the deceased trees of my neighborhood. We had been to Camp Small and to NASA and to the Kilby and Zahradka farms, looking for places to bury trees. We had spent late nights at his home lab here in Silver Spring and hung out on the roof of his campus building after toxic Canadian smoke blew through the region. We had grieved at the foot of the oldest white oak tree in our county.

And in that time—a fraction of a fraction of a blink of an eye in planetary time—more trees had died on Willow Avenue and the year 2023 had turned into the warmest year on record, by a long shot.

"We have to speed up the solutions," Ning said as we still stared at the temperature graph. "That's for sure."

During all our time together, Ning had made it clear to me just how important—how *essential*—the global plan to solve climate change now hinged on removing carbon from the air. Again, because of the lag in the switch to clean energy, scientists at the UN's Intergovernmental Panel on Climate Change said that everything from burying biomass like Ning to paying farmers not to till land to building giant machines to suck the CO_2 out of the air—all these practices together had to equal two hundred million tons of carbon per year by 2030 and a staggering eleven billion tons by 2060 and nineteen billion tons by 2090 when the world, theoretically, would reach net zero greenhouse emissions. Ning had already told me, candidly, that all these approaches were still in their infancy, "like solar energy in the 1970s." And hanging out with him for much of the year had shown me he was right.

"Ning," I said that afternoon, "I look at this temperature graph and then . . . I look at what you're doing."

I paused again for a moment.

"You're knocking on farm doors. You're cold-calling. You're walking up to farms, pretty much on your own, and asking farmers, total strangers, if they'd like to bury small amounts of wood. And meanwhile this."

I tapped the graph. "There's a total disconnect here, don't you think? What you're doing is not nearly enough. Not even close. And will it ever be?"

I had calculated that if Ning buried the full five thousand tons on the Kilby farm, it would constitute less than four seconds' worth of annual carbon emissions for the whole planet.

He didn't grow defensive as much as emphatic. The wood burial work was taking longer than he hoped, for sure, but it *could* be scaled up quickly one day with enough investments and regulatory changes, he said. He reiterated his despair over wasted time and wasted wood. But in the meantime, he reminded me, scientists in Iceland were now using improved methods for injecting CO_2 into water and mineralizing it for permanent storage as rock. And his own daughter, Elisa, had just spent the past summer working as an intern at Lawrence Livermore National Laboratory in California on a project using new machine filters to capture CO_2 in the air. Lots of federal money was going into projects like this from the Inflation Reduction Act, and lots of private money, too.

Still, Ning admitted again, all these methods—and dozens of others for capturing CO_2—were still stuck in the crib of infancy, including his own work.

"So isn't it time," I asked him, "to begin talking more about reflecting sunlight away from the planet? Do you think it's inevitable that we're going to have to try solar geoengineering given the slowness of carbon capture? And, if so, shouldn't we be calling—more publicly and more loudly—for studies on geoengineering?"

Ning, I knew, was never thrilled when I brought up the issue of geoengineering. He had told me privately that he thought it should be studied, yes, but he was pretty skeptical about deployment. He had not signed the open letter issued in February by hundreds of scientists, including James Hansen and Ning's former student Tianle Yuan. That letter called for "robust" funding of solar radiation modification research and experimentation. When we first met last winter and I brought up geoengineering, he had emphatically said there were only three serious legs to the "stool" of climate solutions: mitigation, sequestration, and adaptation. We must mitigate CO_2 emissions by switching to clean energy. We must sequester

CO_2 by pulling it out of the atmosphere and storing it by various means, including as wood or as gas piped into underground caverns. And we must adapt as humans to whatever negative climate impacts occur despite our best efforts at the first two steps.

But a lot had happened in 2023, so I asked Ning the question: "Has anything this year—the wildfire smoke, the temperatures—changed your mind on geoengineering?"

His response came quickly and frustrated me. I had hoped for more nuance—a softening of his position—but he still wasn't there.

"No," he said. "Nothing has changed my mind."

The central idea among geoengineering possibilities, the one most frequently discussed, of using sulfur dioxide or some other aerosol to cool the planet via airplanes in the stratosphere, with humans mimicking a volcano—that idea was one Ning doubted would ever become a safe and viable technique for addressing the climate crisis.

"Study it, yes," he said. "But what I think you'll find is that the earth's systems are so complex, cloud behavior especially, that we just don't know enough about that. We could open a Pandora's box going down that path."

"But haven't we already opened a Pandora's box?" I asked, again pointing to the graph. I mentioned the White House report from late June saying that any consideration of solar radiation modification of the planet should be approached as a risk-versus-risk question. There are risks to engineering the climate, for sure. But compared to what?

"That's a wise approach," Ning said, "but I just think we may not have enough time to fully assess the risks of geoengineering moving forward. Our track record at attempting to modify or control nature has generally not met much success, especially when we do it quickly without enough thought, like with invasive species introduction, for example. We've messed things up. But when we intervene in nature gradually—like the breeding of animals for food and farmwork—humans did that over thousands of years and you can gradually see what works and what doesn't work without risking the whole food system. If you try to do things too suddenly, like with geoengineering, you could starve."

We had reached a familiar impasse between us. Ning thought there wasn't enough time to properly and slowly test, through trial and error, the many challenges that could come with solar geoengineering. And I believed there wasn't enough time for all the small steps, for the slow drip of water, like his work so far, to wash away the great stone of climate change at the speed we now needed.

I wasn't a scientist, of course, and I didn't pretend to be one. But I did trust the scientific prowess of James Hansen, who'd been right at every stage of the climate crisis since the 1980s. He was now engaged in an increasingly high-profile clash of views with leading IPCC climate scientists, saying the scientific body had a history of being too conservative and consistently underestimating the speed and severity of the observed climate impacts so far. Many of those same scientists were now underestimating what was coming soon, Hansen said, and they needed to take geoengineering seriously, even with its risks. I agreed with him.

An hour had gone by since Ning and I took our seats on his porch. The porch itself was now heavily littered with leaves. Ning had used a broom to sweep here before my arrival, but now the place was half-blanketed with yellow poplar leaves and a smattering of oak, sycamore, and beech leaves. Next door, we could hear the acorns of a red oak bombarding his neighbor's backyard deck and roof—*ping, ding, clunk*. Above us, the cloudless blue sky of afternoon was fading to the purple of early evening.

Just then, Ning's son, Luca, nineteen, came onto the porch and introduced himself. He had come home from college that day to help his dad figure out how to use the new truck-weighing contraption Ning had purchased. There on the rear driveway, next to the porch, were the four yellow metallic plates of the scale, spread out to match the four tires of a vehicle. Using gravity sensors and cables connected to a control box, the plates were now calibrated and ready to document the tons of trees soon to be carried up to the Kilby farm.

Luca went back inside, and I prepared to leave. I picked up the piece of paper, the shocking temperature graph, still lying on the table. "This data," I said, standing up and then pointing to the door Luca had just disappeared behind, "no matter how you look at it, it's not good for your son's future and my son's future."

Ning nodded and then did what he always did to exorcise his own fears and channel his own hopes and dreams: he started talking about his work, about putting trees underground. He pulled out his phone before I left, checking the weather app for the third time that afternoon. "The long-range forecast is still good for burial day," he said, obviously trying to lift the mood of the moment. "It's going to be dry, I think. No rain. I think we might be okay."

We had a long handshake that day when I left, a Ning special, with neither of us letting go, standing face-to-face, the unspoken words of respect uttered in volumes between us.

Later that night, preparing for bed, I realized my talk with Ning had been less of a confrontation than a simple search for reassurance on my part. I wanted him to tell me everything was going to be okay, that there really was a plan B. I wanted to hear it directly from someone I knew personally, someone like him, with experience and knowledge and integrity. Our collective global efforts to save the planet weren't happening fast enough, and I wanted Ning to tell me geoengineering would finally save us. But precisely because he was such a high-integrity scientist, he wasn't going to say something he didn't himself believe with high confidence. Others might disagree, but he needed more evidence of probable success before he could consider a strong embrace of this idea.

All of this complicated matters for me, of course. Over the months, I had promoted Ning to a kind of savior status in my mind, I'll admit. Despite his eccentricities and the often unconventional approach of his work, he had become my North Star. Or at least he had found a path and a plan that headed north. For months, that had been reassurance enough for me. Only now I knew his own plan was unlikely to save us either. There was great potential for a partial solution here, but Ning's dream of burying up to three hundred gigatons of wood worldwide this century just seemed impossible. Maybe I knew that all along even as I yielded to Ning's personal and academic charm.

But what I really wanted that October day, I think, was something much more basic, more essential—and equally elusive. I just wanted Ning to tell me there would always be beautiful autumn days like the one today and that the amazing trees on my block—the ones still fighting—were going to stay alive and that my loved ones—my wife, my son, my parents—were going to be okay. There were no assurances, however, in this world of stampeding climate change. Perhaps that was the hardest truth of all. So many bad things had to stop. So many good things—at scale—had to start. And the clock was ticking. We were all living the curse of another, not-so-ancient (it turns out), saying: "May you live in interesting times."

Living in those times now, our time, no one could say for sure just how it was all going to turn out.

Not even Ning Zeng.

11

Burial

November

It was moonless and cold—twenty-eight degrees—when I woke at 4:30 a.m. on November 13, a Monday. I dressed and packed a lunch and grabbed the container of mementos I wanted to take to Ning's tree burial. The professor was bringing a poem to read. Mr. Kilby, the farmer, was going to make a short speech. And I was bringing a few items from my block, from Willow Avenue, to drop into the grave to stay there, covered up, for centuries with the trees.

A day earlier, with permission from the Miller family on Willow, I had pulled a few pieces of bark from the remaining trunk of their once-massive southern red oak. Like everyone who sees this stump—for the first time or the hundredth—I could barely comprehend its size as I walked up. At least eight people, facing outward, could sit along its circumference at one time. I thought again of all the critters it once sheltered in its crown, all gone now: wasps, possums, lizards, tree crickets, katydids, spiders, songbirds, and hundreds more plants, insects, and animals. And I thought of the shade and the cascading sculpture of green—such soothing emotional medicine—it once gave to humans.

On the eve of the burial, I also went to my backyard shed and looked around. Next to an old football, which I grabbed for next week's Thanksgiving game of two-hand touch, I saw what I was looking for: the robin's nest. My son, Sasha, had kept it from a Boy Scout project when he was eleven, lifting it from a honeysuckle bush below the backyard pin oak. I placed the nest inside the ziplock bag with the Miller bark. Outside the shed, that same pin oak was now fifteen years taller and thicker. And it was still masting that autumn. I bent down and began picking up seven acorns among the thousands the squirrels and blue jays had not yet eaten or buried. I counted them. I wanted only seven. I put the acorns in the ziplock with the nest and the Miller bark.

One hundred years from now, whether we've rescued the climate or destroyed it, people will look back and wonder what it was like. Way back in the 2020s, when perilous CO_2 levels and temperatures were not just obvious to scientists but were increasingly felt, in one way or another, by most people everywhere—what was it like to be alive then? How did people make sense of what they were seeing? The newly arriving floods, the dying trees, the heat waves, the smoke? Were people—everyday people—scared? Sad? Angry? Determined to fight? Despondent? How did they talk about it? Did they think about the people of the future?

In microscopic fashion, this book has attempted to answer some of those questions, mostly from the confines of a few square blocks. Here lies a record for future Americans as well as a warning for current ones. This book began when I became utterly fascinated by what I was seeing on my own block. The climate impacts here were screaming a core truth: the fight for our survival is in a phase much later than most of us have realized.

So what were people thinking? What was it like? At least here, among the liberal-leaning citizens of my community, it was never far from most people's thoughts, put there by the screaming headlines and the tumbling trees and the constant water challenges.

So on that cold burial day, as I pulled away at 5:00 a.m., headed for Cecil County and the tree ceremony, I couldn't help but notice my neighbor Dorothy's lights were on downstairs. Why was she up so early on this November morning, there in her home heated by geothermal power? Was she just back from delivering another baby into this uncertain world? Or had she woken in the middle of the night on her own, as she often did, unable to go back to sleep, thinking about her own kids and a future of weather disasters?

The rest of the neighborhood was still dark, bedrooms unlit, everyone sleeping, as I pulled away in the November chill. I could almost feel in that darkness, in the quiet, the resting hopes and dreams and fears of this whole community hovering in the autumn air, allowing me to imagine and recount many of the deeds and aspirations I'd come to see this past year.

I rolled to a stop at Willow and Tulip Avenues. To the left

was my church, that beautiful old stone building with the strangely elevated sidewalk to keep rising water out of the preschool. Soon the church basement, if we could keep the floor from sinking more, would become home to a climate resiliency center for senior citizens and the poor. A possible grant from the county was in the works, and the congregation was eager to see it through.

Past the church, off to my left, I could barely make out the streets that eventually led to my friend Gen Chase's house, where she battled Lyme disease every day in her slow recovery. How many more people, asleep now, would get bitten by a tick that day or soon after and suffer like Gen? Just then, still paused at the stop sign on Willow, I turned on my car brights to better see any surprise crossings of roaming, rutting deer that morning.

Before I turned, heading right on Tulip toward Carroll Avenue, I glanced at the darkened Lawson-Feasley household. No family along Willow Avenue had lost more trees than Kurt and Jill, their corner lot nearly totally deforested. But a handful of baby trees had just been planted—including a beech, an elm, and a pignut hickory—and I wondered if the saplings offered solace to daughter Joanie, still home after the COVID epidemic to save money. Did Joanie, asleep somewhere in that darkened house, still dream from time to time of having children of her own—biological children—despite the world that troubled her so by day? She had turned thirty that year.

On Carroll Avenue now, heading out of town, I passed RS Automotive, the first gas station in America to switch entirely to charging electric cars. In the darkness, an EV taxi was charging as I passed, the solitary driver's face illuminated by the glow of his phone. Mother Teresa had indirectly inspired the creation of this shop, still owned by Depeswar

Doley. Business had increased sharply that year, he told me. My own car cruised by, humming softly, fully charged with electrons from local solar panels.

After Depeswar's, I passed the road leading to Lorig Charkoudian's house, my Maryland state delegate, where I knew *she* was awake, checking early emails ahead of another twelve-hour day. She was always making her vision come true, one step at a time, of a world powered only by wind and solar and batteries. That dream, launched at age twelve, would never die no matter how long it took or what got in her way.

Of course, other people had dreams here, too, long ago. If ghosts roamed the streets of Takoma Park, I imagined they came out in the early-morning hours before light. So in spirit, at least, I could feel the morning presence of those long-ago Anacostan-Piscataway people, still grieving the loss of a place so beautiful that the indigenous word *takoma* translates to "high up" or "near heaven." We can never match the sustainable ways of those early hunters and fishers and traders, but we can try. For two thousand years, native people lived here without wiping out anything, without harming the atmosphere, the very sky above, and causing the trees to get sick and die.

In barely three centuries, we turned that land into scattered farms and then a trolley suburb and then a community, inside a nation, where Congressman Jamie Raskin on Holly Avenue had to try to save the planet by first beating back a wave of US fascism and risking his life.

So people will wonder a hundred years from now, *What was it like?* It was sad and exhilarating and confusing and full of life. And full of goodbyes. Because on every block that morning leaving town, I also passed the cold, dark remains of trees like the Miller Tree, stumps left in the ground or chopped into nothingness, once here and now gone.

A BLANKET OF frost covered the Kilby farm when I arrived an hour and a half later, just before sunrise. The ice was almost glowing, clinging to the rye fields, clinging to everything, a white splendor against the red-orange hues of coming light.

My headlights were still on as I turned down the designated dirt road, bouncing toward the bottomland rendezvous site. In the distance, the Baltimore trees were already there, the wood covered in frost, too, the ice like a faint-white burial sheet tossed over so many jumbled torsos. The trees had been delivered four days earlier and lay stacked haphazardly along an eighty-foot line.

My breath was billowing as I got out of the car. I was the first person to arrive. I walked a circle around the trees and watched alone as the sun peeked over the horizon. I heard a group of crows in the distance—a murder of crows, they're called—cawing in the forest.

Ning and his wife, Annarita, were next to arrive. Ning paced the area nervously, frost gathering on his shoes and pants.

"Can we build a fire to warm up?" Annarita said jokingly. "Any wood around here we can use?"

Ning laughed and relaxed a bit. Annarita was a climate scientist, too, having met her future husband in grad school. She began taking photos of Ning standing atop different tree trunks. He stood on oaks killed by beetles and a maple toppled by high winds and a beech killed by who knows what. The trunks, maybe two hundred in all, were cut into lengths averaging roughly ten to twelve feet. All of them were leafless except for a few covered by the vines and withered leaves of English ivy, with killer and victim ready for burial together.

Around 8:00 a.m., the cold stillness of morning was broken by the sound of the excavator. It moved slowly down the slope, chugging atop tracked feet from a farm building on the hill. Louder it grew, pistons pounding, fronted by a mechanical arm twenty feet long and ready to dig using the might of a four-cylinder diesel engine. A smaller truck, a skid loader, followed behind. It was on tracks, too, and was equipped with mechanical jaws to grip the trees.

Once the heavy equipment arrived, idling next to the trunks, the crew foreman had a final conversation with Ning before commencing the dig. He removed the topsoil from the first section of the grave, scraping away twelve inches with the excavator bucket. The soil was set aside to be returned later, after the burial, when pollinator-friendly grasses and flowers would be seeded here.

Topsoil gone, the real digging began. A member of the crew named Bob jumped into the excavator cab and began clawing methodically into the clay soil, fashioning the deepening walls of the rectangular tomb. He piled the soil on one side of the grave while Hamilton, the driver of the skid loader, busily stacked the trees on the other side, readying them to be lowered. The diesel engines were loud, filling our ears, blotting out all other sounds from the farm and the forest.

The digging went amazingly quickly. Within two hours, Bob had created a grave fifteen feet long, fifteen feet wide, and fourteen feet deep (the first of four sections). Ning periodically lowered a measuring tape into the trench, then signaled for more digging. Twice Ning paused the excavator and climbed into the pit himself to examine the layers of clay more closely, taking notes.

Then, at 10:02 a.m., the first tree was lowered. The excavator stopped digging and swung its arm over to pick up a trunk, pinning the tree against the bucket and expertly delivering the log to a bottom corner of the pit. With that, the first piece of wood—purchased and buried as part of a totally new international carbon market—was laid in place. Mr. Kilby had arrived by then, wearing a gray wool cap, and he and I both shook Ning's hand.

At least fifty more trunks and pieces of trunks were placed in this first burial spot—a quarter of what would be buried that day. The trees were stacked in five horizontal rows, roughly ten per row, with a layer of dirt poured between each row and tamped down by the excavator bucket. The process was noisy and muddy and so completely . . . *industrial*. The vehicles rumbled forward and beeped in reverse. The skid loader zipped around picking up trees with its hydraulic jaws, dropping the logs at precise angles for the excavator bucket. The bucket scooped them up and released them into the grave with a thud so strong you could feel it under your feet. Giant scoops of soil were then dropped back into the grave, covering the trees, and you could feel that, too. After each parcel of dirt landed, the excavator bucket pounded away at it like a boxer, tamping it down. Over and over this happened, the roar of the engines constant. At one point, a V of Canada geese flew overhead, pointed south, the formation low enough for us to normally hear a riot of honking. We heard nothing. A silent V.

After a while, it all just felt like a busy construction site, the ecological romance slipping away as the machines clanged through their work. It was a reminder that, to do this at scale, worldwide, would require an army of trained personnel and a fleet of brute-force machines spread across a thousand spots on almost every continent. It would require money and fuel and big engineering firms. This was no knock on the idea, just another reminder that the globe is a big place to fix.

Yet Ning told me that morning, for the first time, that not all wood sequestration had to involve digging. New ideas were emerging for capturing dead trees along tributaries and then floating them in large numbers to, say, Lake Superior or the Black Sea. The trunks would then be corralled into floating compounds over deep water until the trees grew waterlogged and sank of their own accord to the watery bottom for centuries of sleep (like so many ancient wooden ships still found worldwide). One could imagine a serious quantity of wood being sequestered this way, reviving some of my optimism that tree storage might match Ning's dream as a serious part of the climate rescue. Ning also told me that a new company in Texas had recently buried a few thousand tons of trees in deep dirt, testing different storage methods and soil types while exploring market sales.

But Ning's project, here in Cecil County, was still the first to meet new international standards for verified offsets. A thousand tons had already been sold to Kinnevik, the Swedish firm, and five thousand total tons would be buried here in the spring.

And now, finally, the very first section of this very first grave site was nearly complete. The topmost row of trees had just been put in place, resting uncovered about four feet below the top of the pit. Bob turned off the excavator engine. Hamilton stopped driving the skid loader and killed his engine, too.

Suddenly, the world was still again. Our ears heard more than violent, firing pistons. We heard the mooing of cows and the trill of sparrows in a nearby meadow. The sky was blue. The morning frost had long since burned away. The midday sun was warming the air into the forties.

Nine months earlier at Camp Small, Ning had had the idea of holding a ceremony when this moment finally arrived. Now he plucked his phone from his pocket and glanced at some writing there and called all of us to join him along one side of the burial pit. Mr. Kilby, Annarita, Ning, Bob, and I stood side by side, looking down at the trees as Hamilton filmed from the opposite side of the grave with a phone.

"We are gathered here," Ning began, "for this first wood vault in the whole world to be built to sequester carbon to fight climate change."

He glanced at us witnesses, to his left and his right, and then returned his eyes to the grave.

"First, thanks to the trees," he said, addressing the plants directly. "A gift from nature that helped us with shade, with oxygen, and now after you died, you're still helping with carbon sequestration."

He paused for a moment, as if letting his words sink into the earth. Then he looked at his phone again. "Now I've got a short poem I'll read."

The day before, he had told me three things about this poem. One, it was not from ancient Chinese literature. Two, it was composed instead by Ning with help from ChatGPT—a fact I found strange and a little sad. And three—perhaps blunting the weirdness—Ning really liked how the poem had turned out, and he was afraid he might start crying when he read it. I, meanwhile, had to remind myself that Ning was a scientist, not a poet, and the trees wouldn't care about the overly sweet rhymes.

BURIAL

So Ning began, there on the farm, in the rye field, to read his AI poem:

Beneath the soil, a sacred pledge
Trees entwined, on nature's edge
In burial deep, a climate vow
To sequester carbon, here and now.

And that's as far as he got without fighting back the predicted tears. He began reading the first line of the next stanza, literally saying the words "Whispers in the wind," when an audible breeze passed through the trees of the nearby forest. Ning stopped. He gripped his chin. He blinked. Then he resumed reading.

It wasn't the verse or the soothing breeze making him well up, he later told me. It was the long, long journey he had made. From a start almost thirty years ago in a NASA forest where he first had the idea of burying trees—from that moment forward, he had felt an almost unbearable weight. The weight of responsibility, of knowing that this could and should be done while many people shook their heads. He carried the uncertainty and rejection for years while feeling the added load—the unfathomable tonnage—of more trees dying every year on the planet. If only he could get that first raft of trees in the ground, he thought, the idea would germinate and grow, it would rise up from seed for the whole world to see. And now that moment had come, and he was overwhelmed. Like the heft of those trees landing loudly in the soil that day, a part of Ning's burden was coming off his shoulders.

Halfway through the poem, he paused a second time, putting his hand on his chin again, waiting, then rallying with an even shakier voice for the last stanza:

Nature's ballet, a carbon dance
Buried hopes, a last chance
In soil's embrace, a promise made
For a greener future, in quiet shade.

Right after "In soil's embrace," he stopped a final time and put his hand on his heart, covering his chest. We weren't sure he was going to make it this time. He left his hand there, on his heart, shaking his head back and forth, eyes watering, still pausing, until he finally got through it, reading those final words.

Nobody wanted to follow him after that. Annarita said nothing. Mr. Kilby, clearly affected, spoke briefly: "I'm not sure what I would say. It's just very meaningful to see Ning so moved by what he's doing, and you know I feel very blessed to have him so interested in doing this project. So thank you."

I was the last to go. I pulled out the bark from the Miller Tree on Willow Avenue and quickly told its story. I scattered the bark, three pieces, over the deceased trees, the trunks prone and mute below, receiving the offering. Next I dropped Sasha's bird's nest into the grave, representing—to me—the fragile realms of nature Ning was trying to save.

And then the acorns. Pin oak acorns are characteristically small and round, not oblong, each one like a tiny Earth. I held seven in my hand to represent all seven continents on the planet. I counted them out, one by one, releasing them slowly onto the noble trees below.

Epilogue

December Endings

I was unloading Christmas trees from Mike Tabor's box truck when he asked me, "Have you heard the story about the tree farmer and the Messiah?"

It was mid-December, a mild day, almost shirtsleeve weather, but nothing seemed to dampen the holiday mood. People across Takoma Park kept calling and coming by Mike's house to pick up Christmas trees. The "Tuba Santa" had been playing at the downtown gazebo for days, and red bows hung from many city streetlamps.

Every year, I try to buy my tree from Mike, my dear friend, and his amazing wife, Esther. He's eighty-one, and she's seventy-five—and they still own and work a sixty-acre farm in southern Pennsylvania. They use their home in Takoma Park, on Erie Avenue, as their base for the several days per week they sell produce and honey and apple juice at DC-area farmers markets. Perhaps surprisingly, they're Jewish, and a Christmas tree has never been inside their house. Mike built half of that house himself, by the way, using backyard clay for walls and straw bales for insulation.

"Please, no stories about the Messiah!" Esther blurted out just then. "When you talk, Michael, you stop working. Please focus." She was measuring and tagging the trees he handed her from the truck—Fraser firs, white pines, blue spruces. Mike even planted a few concolor firs each year just because he loved the way they smelled like fresh citrus. After the trees were tagged, I carried them to the side of the house for customers to sort through.

Mike started telling the Messiah tree story anyway until Esther quickly cut him off. "Michael! Seriously! Don't talk."

That's what it was like in the company of these two, an utterly loving, adorable, bickering couple who instead of casually finishing each other's sentences more often said, "Can I finish my sentence, *please*?"

I buy my holiday trees from them because they grow "low carbon" trees on a sustainable farm. They plant the trees on a patch of marginal land not good for much else. They don't use chemical herbicides or fertilizers, so the trees grow slowly while absorbing carbon on otherwise fallow land. Those trees then shine radiantly, strung with lights and ornaments, in the windows of many Takoma Park homes during December, ours included. After the holiday, I toss our tree in the far backyard to serve as winter shelter for birds and other critters.

Beth and I try to weigh the climate consequences of our

individual choices whenever we can. We are vegetarians, we fly as little as possible, we have solar power, and we buy our Christmas trees from Mike and Esther. But none of this really matters. We do it because it's right, not because it changes the big picture.

When people ask me, the climate activist, what individual steps they can take to fight global warming, I tell them, "Step one is stop being an individual. Join a *group* of other individuals instead and become part of the climate movement."

Of course, it's good to make changes at home. But the oil and gas and coal companies *want* us to focus on our carbon footprint. They want us to believe that global warming is a matter of individual choices—not about their tobacco-esque lies and public deceptions. *If you don't like the rising seas and dying trees, make different choices at home, bro.*

But if you're asking that same question after reading this book—*What can I do?*—my answer is simple: find an organization fighting climate change in your area and develop a relationship with that group as a volunteer or a donor or both. A Google search will turn up local chapters of Interfaith Power & Light and the Sierra Club and college climate groups in most states. There's even a new climate organization specifically for baby boomers called Third Act. And regional groups like the Chesapeake Climate Action Network exist in most parts of the country, augmenting the work of well-known national groups. We did not, as a country, end racial segregation and stop child labor practices and phase out DDT through voluntary individual choices. Those victories required national movements of people demanding statutory change. The climate movement, likewise, needs more citizen activists to hasten the switch to clean power and finally abolish fossil fuels *by law*.

In early 2023, when I started writing this book, I did not know the year would be the warmest on record—by a lot. Every year has a shot at the title when the warming trend is so strong, of course. Maybe 2025 will be the new record. But the heat of 2023 was a run-and-jump leap beyond everything before, a whopping 1.35 degrees Celsius (2.43 degrees Fahrenheit) above preindustrial levels and beating the previous record year of 2016 by more than a quarter of a degree Fahrenheit, a margin that astounds climatologists. This new heat has made the world so scrambled, with such extreme impacts, that—as I said at the outset of this story—a person can now throw a dart at a spinning, lacquered globe and wherever that dart lands, right at that tiny place, you can write a whole book about the overt climate disruptions happening there. That's as real as it gets.

By definition, recapping the regional highlights of that warmest year on record is a whipsawing exercise. It began in DC with the third-warmest winter here, followed by that weirdly cool spring triggered in part by high-elevation, sun-blocking smoke from Canadian wildfires. Then came the choking low-elevation smoke of June and early July, never before seen here. Then a late, blistering heat wave came in September, with the mercury hitting one hundred degrees at Washington's Dulles International Airport for the first time ever in September. Then the tenth-driest October—before the rain came roaring back in late fall with a vengeance. From late November to late December, we saw nearly seven and a half inches of rain in DC, missing the record for that period by a tenth of an inch. Through it all, most of the DC region maintained its longest snow drought on record, reaching 656 days at Dulles without one inch of snow in a calendar day. The snow deficit was similar in locations from New York's Central Park to counties in southern Virginia.

At the global level, as I wrote this book, the year 2023 began with shockingly low sea ice in Antarctica, a new record. Then, among other disasters, came the worst flooding in 140 years in Beijing and the Maui wildfire in August that killed more than one hundred people. In September, when it was still wintertime in Brazil, an impossible heat wave drove the temperature to 111 degrees Fahrenheit, just one degree short—in winter—of the country's all-time record for *any* season.

Years ago, I saw Al Gore give a version of his famous slideshow to an international audience in Copenhagen, Denmark. After heartbreaking slides of melting glaciers and flooded coastal villages and parched savannas, he ended with a photo of a completely burned-down forest in California, the ash-covered ground and blackened trees stretching to the horizon.

"This," he said, pointing to the screen and then looking back at the audience. "This is not who we are. This is not us. This cannot be our destiny. We are greater than this."

I have always remembered those words and have always believed them, despite everything. This is not us.

So who are we, then? I don't know exactly, but every December in Takoma Park, I catch a glimpse of it, I think, at my church's very fine choral presentation to kick off the Advent season of Christmas. Maybe that year, 2023, we all needed it more than normal as we filed into the sanctuary on Sunday, December 3—not just the church regulars but, per tradition, half of the town seemed to come: the religious and nonreligious, people of every age, every race, parents with kids, everyone packing the balconies, filling every pew, standing room only. Our excellent music director and choir, combined with the local Washington Adventist University choir, plus an orchestra made up of local players and music majors from the university—they all proceeded that day to create something I can't fully describe.

Professional singers who've performed at the Kennedy Center and Carnegie Hall rave about the mystery that is the fine acoustics of our modest sanctuary, with its one-hundred-year-old wooden floors and domed ceiling. And by the time all twelve movements of Antonio Vivaldi's *Gloria* had been performed that December day, and the cellos had soared and the classical, almost operatic solos had been sung, and the full climax had arrived with every voice swelling—sopranos, altos, tenors, basses—and every instrument playing—trumpets, violins, oboe—when all that had happened, there was nothing separating any two people in that sanctuary. Everyone and everything came together, the music melding with the hand-painted banners on the sanctuary wall, banners hung amid a sea of red poinsettias and strands of white lights, saying: HOPE. PEACE. JOY.

This, I think, is who we are. Deep down.

But we get lost. A lot.

For Beth and me, the uplifting choir and the tree from Mike Tabor were not the only things bringing a needed boost to our moods that season. Six days before Christmas, two gifts arrived: a lovely bald cypress and a swamp white oak, both delivered by Casey Trees at last. Within an hour, the six-foot-tall cypress was settled into soil in the backyard, deer guard around its two-inch-thick trunk, ready to endure whatever soggy weather climate change had in store for us. (If the baby pin oak, raised from an acorn twenty-five feet away, could hang on, we'd have quite a backyard understory in coming years.) Meanwhile, in the front yard, the swamp oak was our best attempt at climate adaptation right there on Willow Avenue, a few feet from the curb. I offered a Christmas wish out loud as the tree went into the ground: "May you grow old and beautiful on this street for one hundred and fifty years, always strong and fit."

But a wish, as they say, is not a plan. And a plan is what you need when you're lost. Back at the church, down in the basement gym next to the preschool, I had spent years teaching my son and other boys—and myself—what to do when you get *really* lost. That wilderness survival class in Boy Scouts is what came back to me a year ago in my backyard as I sat under my struggling adult pin oak wondering how to navigate out of the paralysis I felt over climate change. Stop, think, observe, plan: STOP. That's how you get out of the wilderness. And I had certainly followed the first three steps of that method over the last year. It had been a long pause of a year where I thought hard about the pathways that got us lost on climate, and I observed everything and everyone around me for a possible way out.

But the *P* was the hardest part of that method—the plan. The clean-energy transition was well underway worldwide, yes, with revolutionary effect. In 2023 alone, a gigawatt of solar power was installed every day across the globe for much of the year. But even this, again, was not enough given the ticking clock. We started too late. So the global plan now was of the hope-for-the-best variety while we hoarded as much carbon as possible underground as rock, as gas in subterranean caverns, as trees everywhere.

And beyond that? Geoengineering? Yes, it's playing god with the planet—and we've been doing it for centuries now. In the short term, it would be helpful to maximize ways to reflect sunlight away from the planet with white-painted roofs and mirrors on land and ships at sea creating brighter clouds. But there's a reason Dr. James Hansen and others are prioritizing the conversation around incrementally adding sulfur dioxide to the stratosphere. Done carefully, it would start cooling the planet almost instantly and at relatively low expense.

And here's what I'll add: I've been pretty good at predicting a few big things in my life. In the early 2000s, I left a great journalism job and reordered my career into that of a nonprofit leader because I saw it coming. I saw climate change absolutely dominating our lives for the rest of this century, all of us, on every single dot on the planet. Then, in 2003, as part of that awakening, I actually wrote a book—*Bayou Farewell*—that predicted the warming-enhanced arrival of Hurricane Katrina a full two years before it happened.

Now something else is crystal clear in my mind: in the next decade, as climate impacts inevitably worsen from the sheer momentum of the emissions already in the atmosphere, political interest in geoengineering will grow rapidly worldwide. The first rule of politics, after all, is to stay in power. What will happen when alarmed and potentially food-insecure masses gather outside the parliaments of Europe and the presidential gates in Beijing and outside the White House—all demanding answers, demanding a rescue? For me, it's not hard to see leaders reaching for some kind of novel lifeboat.

I think it's okay that people disagree on whether solar radiation modification can be done effectively and safely in the stratosphere. That doesn't matter as much as the need to study it now and research our different options and conduct limited experiments to rule out the bad ideas while setting aside the stronger approaches for later consideration (using a risk-versus-risk assessment approach ahead of any final international decision to deploy). Simply put, in ten or twenty years, when climate change is unavoidably worse than it is now—and it's bad now—our plan should be to have more options than just enduring the deep suffering of a burning planet.

For some people, even the discussion of geoengineering makes the situation feel hopeless. "If that's where we are right now, it's already too late," they say.

I disagree. I recall my pastor Mark Harper saying, "If we have half a chance to save humanity and the earth, we should consider all ideas, including this one."

And in the broader, uphill fight to rescue the world from runaway heat, I recall a quote often attributed to Dr. Martin Luther King Jr., though it may be apocryphal. During a dark moment of the Civil Rights Movement, after a riot, King was supposedly asked how he maintained hope. He paused and said, "Even if I knew that tomorrow the whole world would go to pieces, I would still plant my apple tree today."

Which brings me back to Mike Tabor, the Christmas tree farmer in Takoma Park. On that too-warm December day, as he and Esther and I unloaded that truckful of trees, Mike finally finished his story about the tree farmer and the Messiah. It helps to know, as background, that Mike has spent fifty years making his own farm succeed against extraordinary odds. Don't get him started, for instance, about the climate-change troubles he's seen in that half century at Licking Creek Bend Farm. Don't ask about the heat and the weird frosts and the flood that almost took the life of one of his farm staffers four years ago. Or the Lyme disease he's had twice and thinks he may still have. Or, in Takoma Park, the trees he and Esther have seen decline in their neck of the neighborhood. They lost a giant red oak in their backyard in 2019, but instead of a stump, Esther asked the tree-removal crew to oddly leave twenty-five feet of the trunk still standing.

That's so she can look out her second-story window and, even up there, pretend the once-beautiful tree is still alive.

Despite all this, Mike announced there in the truck that day, during my visit, that he planned to keep working six days per week for another ten years, into his nineties. He just didn't want to stop farming, he said. Esther rolled her eyes at me, but I think she took him seriously.

And then, with Esther finally yielding, Mike told his story:

"So there was this tree farmer," he said, "and he's planting a tree one day, and an old man walks up, an older farmer. And the man says to the younger farmer, 'No matter what happens, don't stop digging until you've finished planting that tree.'

"'I know,' the young farmer says.

"'No, really,' the older farmer says. 'Don't stop.'

"'I know.'

"'Even if the Messiah himself shows up . . .'

"'The Messiah himself?'

"'Even if the Messiah shows up, keep digging, and don't even greet the Messiah until that beautiful tree is in the ground. It's just too important. Do you understand? Too important.'

"'I understand,' the young farmer said."

And that was Mike's story.

Postscript

As this book goes to press in November 2024, nearly a year has passed since Ning Zeng buried one hundred tons of trees on farmland in Cecil County, Maryland. During that time, hard-at-work climate scientists worldwide have yet to fully explain the accelerated, record-smashing, month-by-month temperature rise globally that began in June 2023 and has continued to the present. We've obviously crossed some kind of threshold, possibly related to the declining ability of the world's forests to absorb carbon dioxide as they are battered by climate extremes. According to one temperature measurement—from the European Union's Copernicus Climate Change Service—the planet actually reached a sustained temperature of 1.5 degrees °C above preindustrial levels in early 2024. If the trend continues, the planet could permanently cross that threshold in the next few years, reaching a level scientists continue to warn will have staggering consequences for all life on Earth.

Donald Trump's reelection is a clear setback for efforts to address this emergency. Still, the clean energy revolution—

as I argue throughout this book—cannot be stopped. A hostile White House can slow it down, but the transformation will continue in this country in part because nearly half of all the clean-energy jobs and manufacturing investments created by the federal Inflation Reduction Act of 2022 are accruing in the seven swing states that now influence Presidential election outcomes: Georgia, Nevada, Arizona, North Carolina, Pennsylvania, Michigan, and Wisconsin. Those states are unlikely to give up these gains easily. Of course, new challenges will always be with us on the path to a carbon-free world, including the sudden threat of fast-spreading, energy-hungry data centers and the hell-bent drive of the American fossil fuel industry to export liquefied natural gas to the rest of the world even as "degasification" ambitions grow in the United States.

This combined picture—temperatures rising faster than the steady-but-still-not-enough deployment of clean energy—means that storing carbon in the ground has never been more important. Which makes the uncertainty over Ning's buried trees in Cecil County even more tragic.

Those hardwoods and pines from the streets of Baltimore were meant to stay locked in the clay soil for a thousand years or more on that farm. But a wrinkle in Maryland state regulatory law has thrown a wrench into the plan.

Shockingly, unthinkably, the Maryland Department of the Environment (MDE) is now telling Ning and farmer Bill Kilby to dig up all the trees. The agency actually wants the nine-hundred-square-foot tomb opened up, the dirt cast aside, and the trees lifted and moved to another site. Whether this actually occurs is still an uncertainty as of this writing.

So, what the hell happened? For me, it's hard to tell how much of the situation is an honest misunderstanding of staggering proportions and how much is Ning subscribing to the "move fast and break things" ethos of innovators in places like Silicon Valley and beyond. I've reviewed documents and spoken with many of the major players in this emerging regulatory conflict, and a misunderstanding seems to be the culprit. In the end, Ning and Mr. Kilby never intended to move forward without what they believed was firm approval from local consultants and regulators. That much is absolutely certain.

Yet in the summer of 2023, before the trees were buried, MDE officials did tell Ning that burying trees per his plan would qualify as waste material, requiring a very onerous landfill permit. Ning informed them that he wasn't creating a landfill, but instead a "de-nitrification" system permissible through the US Department of Agriculture's Natural Resources Conservation Service. A USDA representative of the Cecil County field office had assured Ning that as long as the tree project was smaller than 5,000 square feet of surface area and ringed by a deep trench of wood chips to help neutralize nitrogen from the farm's fertilizer runoff—a plus for the environment—then a state-level permit would not be necessary. Such systems are also permissible on farm-conservation easements in the state.

Ning's confidence in this approach—along with that of his consulting engineers at Wetlands Studies and Solutions, Inc.—can be measured by the utter transparency of the operation. Three weeks before the burial, Ning and Mr. Kilby held a community-outreach event at the farm with neighbors, local officials, and USDA soil conservationists. Then, during the week of the burial, in November 2023, Ning invited a reporter and key members of the Maryland General Assembly and officials from the US Department of Energy to view the worksite. They all came.

But in the winter of 2024, as a bipartisan pair of state lawmakers introduced legislation to explicitly declare that wood burial for carbon storage, done properly, would not need conventional landfill permitting, MDE opposed the bill, asked the lawmakers to withdraw the legislation, and openly disputed the Ning, Kilby, et. al. approach to wood burial at the Cecil County site. The trees had to be dug up, MDE said, and reburied on non-conservation-easement land using a landfill permit, just as if the trees were a conventional waste product, which is of course absurd.

Ning and Mr. Kilby have argued passionately against the removal, saying exhuming the trees would have a bigger environmental impact than leaving them there, and asking MDE officials to at least permit this as a small research project—one that, within a few years, will erase the agency's concern of possible underground fires and chemical leaching into the groundwater. Those dangers are all but impossible given the use of whole logs in a near airtight and oxygen-free environment. But the agency is not budging, and something of a standoff is in place.

Ning is hopeful that legislation can pass in the 2025 Maryland General Assembly session to give "grandfather" approval to the Kilby wood vault and create more pilot projects across the state without the burden of unnecessary landfill status. It's unclear how this will turn out, but everyone outside of a small group of regulators at MDE seems to agree that ripping the resting trees from the soil is just plain madness.

Meanwhile, in anticipation of new legislation in Maryland, Ning is working to design a burial project on a farm in Charles County, Maryland, and another in West Baltimore, where an abandoned rock quarry could serve as a potential tree tomb, covered by the excavated soil of a nearby Amtrak tunnel project.

On another front, after a long wait as a finalist, Ning did not get the million-dollar XPRIZE from entrepreneur Elon Musk in April 2024. He did receive a $50,000 grant from the US Department of Energy to work on burial projects in Maryland, Montana, and, potentially, West Virginia.

The story of the Kilby Farm, meanwhile, further illustrates the barriers and regulatory delays facing innovators like Ning, who want to implement climate-saving technologies at scale while there's still time.

"I will never stop," Ning told me as I was writing this postscript. "I intend to live to see—in my lifetime—a gigaton of carbon buried as trees around the world every year."

That would be the equivalent of removing the present-day emissions of Britain, Spain, and Australia every year from the earth's carbon-pollution budget. It's a start.

"I will live to see that," Ning repeatedly told me. "I won't stop until it happens."

ACKNOWLEDGMENTS

Writing a book about your immediate neighborhood is a terrifying idea. Screw up, and the complaints to the publisher will be nothing compared to the stern word at the grocery store or the disapproving glance across the church pew or the impromptu feedback during a morning walk. Which is why I'm so grateful to so many people in Takoma Park who went out of their way to help me get things right—as much as humanly possible—on these pages. Old neighbors and new friends across the city supplied facts and surprising points of view on how climate change is affecting their lives. From these accounts—recorded in backyards, in living rooms, and among the magnificent street-side trees both alive and deceased—I was able to piece together a collective narrative of what is happening to us in this one tiny spot on a changing planet.

So, thank you first goes to my immediate neighbors on Willow Avenue, many of whom offered edits on early drafts of the book: Lisa and Dave Miller, Dorothy Lee, Kurt Lawson and

Jill Feasley, Joanie Lawson, Jim Witkin and Nancy Flickenger, Ashley and Audrian Flory, Kathie Hart, Hemakshi Gordy, Michele Kurtz, Fred Pinkney, and Denny May. Thanks also to non-Willow neighbors: my pastor, Mark Harper, Bob Gibson, Catherine Varchaver, Mike Tabor, Esther Siegel, Lorig Charkoudian, and of course, Congressman Jamie Raskin.

Additional gratitude goes to the Takoma Park Public Works Department, especially Deputy Director Ian Chamberlain, who spent hours explaining to me—by phone and in the field—the "hydrological contours" of a city struggling with bigger storms while implementing increasingly creative solutions. Thanks also to Ian's boss, Daryl Braithwaite, and to urban forest manager Marty Frye.

Arborists and entomologists of many stripes—commercial, government, academic—were interviewed for this book. Special thanks to Chris Larkin, senior arborist at Bartlett Tree Experts, who knows more about the trees of Takoma Park than anyone. And thanks to Chad Rigsby, an insect ecologist at Bartlett; and Ken Raffa, emeritus professor of entomology at the University of Wisconsin, who checked many of my bug facts. Shaun Preston, yard manager for Baltimore's Camp Small—aka the "stump dump" for downed trees—was also a big help. As was my friend Barbara Briggs of the Sierra Club's Beyond Gas initiative.

Thanks also to the board and staff of the Chesapeake Climate Action Network, especially Dave Goodrich and Jamie DeMarco, and my assistants Riley Pfaff, Hadley Dzuray, and Diana Menendez.

Special compliments to Jennifer Lyons, my extraordinary literary agent of twenty-five years, for her Job-like patience—and to my editor, Pete Wolverton, at St. Martin's Press. Pete came to love the trees of my neighborhood as much as I did and offered keen edits to allow their story to take full command, rightfully, of this narrative.

Finally, thanks to Dr. Ning Zeng of the University of Maryland, the most passionate climate scientist I ever expect to meet. During many improbable adventures together, spanning a record-hot year, he taught me more than just how to store carbon molecules safely underground in the form of buried trees. He taught me to believe, all over again, that we can still get out of this mess.

Of course, I could not have written this book without the constant support of my family, especially my dad, Wayne Tidwell, and my son, Sasha Tidwell. And then there's Beth, my dear wife, who wakes up every morning to life with a climate-obsessed writer and activist. Only she knows the grace and calm she brings to our inner world as the storms of a fragile, beautiful planet blow outside. *Te amo*, Beth.

INDEX

Aboudou, Wabi, 171, 187
acorns, *1*, 3, 9, 11, 51, 117, 155, 213–8, 242, 252
activism, climate, 25–39, 79–82, 93, 142–3, 151, 159–64, 255
AERONET (Aerosol Robotic Network), 103–5, 108–10
agriculture
 climate change's impact on, 188–91
 regenerative, 5, 46, 55–6, 71
ambrosia beetles, 16–21, 23, 48–50, 62, 174, 216, 232
Anacostan-Piscataway people, 11, 125, 245
animals/insects
 climate change's impact on, 2–3, 20–1, 51, 155–6, 174–80, 215, 237
 as tree killers, 16–21, 23, 48–50, 62, 65, 174, 216, 223, 232
 trees' symbiosis with, 3, 9–10, 129, 242
Antarctica, 2, 14, 25–7, 48, 104, 187, 257
asthma, 137
Atlantic Coast Pipeline, 31–2, 201

atmosphere/stratosphere,
 engineering of, 27, 72, 95–9, 103–11, 149, 236
 cost of, 112–3, 260
 ethical considerations on, 114–21, 126, 167–8, 259
 ozone layer impacted by, 114–5
 treaties/agreements regarding, 112–4, 166, 168

Babesia microti bacteria, 180
Bay Journal, 50–1
Bayou Farewell (Tidwell), 260
beetles, tree destruction by, 16–21, 23, 48–50, 62, 65, 174, 216, 223, 232
berm systems, flood, 75, 79, 82–5
Biden, Joe, 32, 36–7, 125
biochar, 55
bio-energy with carbon capture and storage, 56
Bond, Julian, 29–30
Borrelia burgdorferi bacteria, 180. *See also* Lyme disease
Braithwaite, Daryl, 23, 53, 83
Briggs, Barbara, 135–40, 217

brown, adrienne maree, 162
Brown, Autumn, 162
Budyko, Mikhail, 98–9
Buff, Jesse, 193–5

Camp Small, Maryland, *43,* 60–5, 67–8, 148, 154, 210–1, 218, 221–9, 233, 250
Canada, wildfires in, 131–2, 135, 146–53, 157, 160, 177, 233, 236, 256
Capalongo, Dillon, 218–28
carbon dioxide, 2, 4, 70–1
  legislation capping, 34, 255
  offset programs for, 184, 224, 248–9
  plant growth expedited by, 128, 192
  target goals for, 46–7, 54, 56–7, 234
carbon dioxide removal, 36
  via direct air capture/storage, 5, 46, 56, 71–2, 108, 234–6
  via injection into water, 235
  using photosynthesis, 56, 128
  via regenerative agriculture, 5, 46, 55–6, 71
  using rocks, 5, 55–6, 235, 259
  time as factor in, 27, 38–9, 46–7, 54–7, 206, 213, 232–40, 259
  via tree burial in ground, 52–3, 58–68, 76, 100, 148, 151–4, 181–6, 209–13, 218–40, 241–2, 246–52
  via tree burial in water, 249
Carbon Lockdown, 71, 182
Cardillo, Gary, 77, 120
cars
  electric, 4, 20, 27, 35, 71, 88, 99, 119, 126, 141–5
  gas-powered, 53, 141–2, 144–5, 163, 183, 244
Carson, Rachel, 22
CCAN. *See* Chesapeake Climate Action Network
Centers for Disease Control and Prevention, 178
Chamberlain, Ian, 82–8, 126

Charkoudian, Lorig, 197–208, 245
Chase, Gen, 244
cherry trees, 69–71, 73, 90–2, 94
Chesapeake Climate Action Network (CCAN), 27–39, 80, 178, 201, 255
clean energy, 168
  via electric cars, 4, 20, 27, 35, 71, 88, 99, 119, 126, 141–5
  via heat pumps, 33, 71, 124–7
  infrastructure for, 141–5
  legislation supporting, 4, 27–39, 66, 71, 79–82, 127, 201–2, 206, 211, 235, 255
  nuclear-based, 36
  solar-based, 4, 20, 27, 32–5, 37, 71, 88, 125, 127, 134, 146–8, 186, 190, 203, 206, 245, 259
  transition time/plan for, 38–9, 46–7, 57, 108, 125–7, 235, 259
  wind-based, 27, 34–7, 80, 125, 127, 146–8, 190, 197–208, 245
Clean Energy Jobs Act, 202
climate change. *See also* solutions to climate change; trees; weather changes
  activism, 25–39, 79–82, 93, 142–3, 151, 159–64, 255
  health impacts of, 2–3, 20–1, 44–6, 137, 155, 160–4, 174–80, 203, 244, 261
  legislation/mandates addressing, 4, 27–39, 66, 71, 79–82, 125, 127, 149, 166–8, 201–2, 206, 211, 235, 255
  speed of, 25–7, 38–9, 46–7, 54–7, 153, 213, 232–40
  wildfires tied to, 5, 48, 51, 65–6, 109, 131–2, 135, 146–53, 157, 160, 177, 210–1, 233, 236, 242, 256–7
  wildlife impacted by, 2–3, 20–1, 51, 174–80, 237
cloud engineering. *See also* atmosphere/stratosphere, engineering of

INDEX

brightening of clouds as, 46, 103, 107, 113, 259
via chemical modification, 29, 98, 236
coal, 2, 30–7, 106, 109, 123, 126, 128, 255. *See also* fossil fuels
Community Choice Energy Act, 202
Community Mediation Maryland, 199
Congress, climate action by, 4, 27–39, 79–82, 116, 127, 149
Conowingo Dam, Maryland, 229–30
Constellation Energy, 229
coral reefs, 14, 187
corn-burning stoves, 123–4
COVID pandemic, 44, 78, 159, 180, 189, 192, 195, 244
Crutzen, Paul, 114

Dawes, Melvin, 144
deer, 2, 155–6, 158, 164–5, 179, 215, 244
DeMarco, Jamie, 37
Doherty, Sarah, 72
Doley, Depeswar, 141–5, 244–5
Doley, Evaline, 141
Doley, Teresa, 141–2
Dominion Energy, 31–2, 172–4, 201–5
driftwood, 223–4, 229–30
droughts, 9, 12–3, 18, 20, 27, 38, 47–8, 51, 88, 146, 153, 156–7, 164, 210, 215–7
snow, 101, 256

Ehrlich, Robert, Jr., 80
electric cars, 4, 20, 27, 35, 71, 88, 99, 119, 126
infrastructure for, 141–5
Ellen, Terry, 178
The End of Nature (McKibben), 27
English ivy, 191–7
EPA (Environmental Protection Agency), 136, 138, 179

Escobar, Jazmine, 218–28
ethanol, 16–7, 216
ethical considerations, 114–21, 126, 167–8, 259

Feasley, Jill, 163, 244
FEMA, 162
flooding, 3, 27, 38–9, 48, 70, 72, 164, 229, 242, 257
and prevention measures, 2, 75, 79, 82–8, 126
in Takoma Park, 2, 13–4, 73–9, 80, 82–9, 92–4, 108, 199–200
Flory, Ashley, 171, 177–8
Flory, Audrian, 178
fossil fuels, 2, 25–33, 35–7, 53–4, 80, 123–8, 190
cars powered by, 141–5
and gas leaks, 129, 135–40
legislation/mandates on, 34, 255
sulfur dioxide release by, 105–6, 108–9, 111, 114
transition away from, 38–9, 201, 205
fracking, 30–1, 36, 80
Franklin, Ben, 98
fungi, 12, 15, 17, 133

gas. *See* fossil fuels; methane gas
Gates, Bill, 113
General Motors, 145
geoengineering. *See also specific categories*
carbon dioxide removal category of, 5, 27, 36, 38–9, 46–7, 52–68, 71–2, 76, 100, 108, 128, 148, 151–4, 181–6, 206, 209–13, 218–40, 241–2, 246–52, 259
ethical considerations on, 114–21, 126, 167–8, 259
solar radiation modification category of, 5, 27, 29, 46, 55, 72, 95–9, 103–21, 126, 149, 166–8, 206, 235–7, 259–60
geothermal energy, 132, 134, 176, 200
Gordy, Michael, 10, 70, 171

INDEX

Gore, Al, 257
Great Barrier Reef, Australia, 2, 107
Greenberger, Scott, 10, 15–6, 51, 76
"greenhouse effect," 27
Greenland, 47, 102, 167
Griffith, Saul, 37
Guterres, António, 4, 54

Hamer, Fannie Lou, 118
Hansen, James, 39, 72, 96, 98, 107, 204, 235, 237, 259
Harper, Mark, 116–21, 261
Hart, Kathie, 132, 188–9
Hart, Laird, 132, 188–9
Harvard University, 5, 113
health, climate change impacting, 2–3, 20–1, 44–6, 137, 155, 160–4, 174–80, 203, 244, 261
heat pumps, 33, 71, 124–7
Hitt, Mary Anne, 37
Hogan, Larry, 143
Howerton, Keith, 156
How to Survive the End of the World (Brown/brown), 162
hurricanes, 3, 14, 18, 27, 39, 51, 66, 72, 107, 170, 187, 260

ice caps/sheets, melting of, 2, 5, 14, 25–7, 47, 48, 102, 113, 153, 187, 257
Inflation Reduction Act of 2022, 36, 66, 71, 127, 235
Intergovernmental Panel on Climate Change, UN (IPCC), 51, 56–7, 66, 234, 237
ivy, invasive, 191–7, 232

January 6 insurrection, 80–1, 207
Jemsek, Joseph, 179
Johnson, Lyndon B., 28, 35, 98, 107

Katrina, Hurricane, 14, 72, 260
Keith, David, 113–5
Keystone XL pipeline, 29, 32
Kilby, Bill, farm of, 185, *209*, 212–3, 218–25, 233–4, 238, 241, 246–52

King, Martin Luther, Jr., 78, 261
Kinnevik, 184, 249
Kurtz, Michele, 10, 15–6, 51, 76

land-based solar reflection
  via mirrors, 112–3, 259
  via painted surfaces, 113, 259
Larkin, Chris, 20, 23, 52
Lawrence Livermore National Laboratory, 235
Lawson, Joanie, 158–65, 168, 244
Lawson, Kurt, 73–9, 133, 163, 244
Lee, David, 132
Lee, Dorothy, 131–4, 140–1, 159, 174–7, 180, *181,* 187, 189–91, 243
Lee, Kenneth Allan, Jr., 172
legislation/policy, climate-based, 4, 27–39, 66, 71, 79–82, 125, 127, 149, 166–8, 201–2, 206, 211, 235, 255
Leventhal, George, 215
Limpert, Bill, 31–2
Limpert, Lynn, 31–2
Linden Oak, Bethesda, 181–5
Lopez, Alfonso, 37
Lyme disease, 2–3, 21, 44–6, 155, 174–9, 203, 244, 261
  vaccine for, 180

Manchin, Joe, 36
marine cloud brightening, 46, 103, 107, 113, 259
Maryland Offshore Wind Act of 2013, 34
May, Denny, 188–9
McGill University, Montreal, 66
McKibben, Bill, 27, 46
Menendez, Arsenio, 104–5
methane gas, 2, 34, 36, 71, 80, 106, 124–8, 153, 220, 236, 255, 259. *See also* fossil fuels
  cars powered by, 53, 141–2, 144–5, 163, 183, 244
  pipes/pipelines, 4, 30–2, 129, 135–40, 201
Miller, Dave, 9, 11–2, 14, 17–9, 241
Miller, Lisa, 9, 11–2, 14, 17–9, 241

INDEX

Miller Tree, 9, 10, 17–20, 43–4, 51–2, 76, 241–2, 246, 252
Miller, Wesley, 9, 12, 18
Miracle Ridge, Virginia, 31–2
mirrors, solar reflection using, 112–3, 259
Moore, Wes, 185
Mount Pinatubo, Philippines, 95–7, 107, 131
Mount Tambora, Indonesia, 98
Moyer, Lezetta, 19
Moyer, Lin, 19
mulch, 50, 61–4, 189–90
 and fire risk, 210–1
Musk, Elon, 60, 71, 112

NASA Goddard Space Flight Center, Maryland, 39, 59, 98–112, 147–8, 204, 206, 233, 251
National Gardening Association, 189
National Oceanic and Atmospheric Administration (NOAA), 28, 116, 187
The Nature of Oaks (Tallamy), 129
negative emissions, 5, 53–7, 60, 66, 71. *See also* carbon dioxide removal
Neill, Pat, 171, 187
New York Times, 129
nuclear energy, 36

oak trees, 181–5. *See also* trees; *specific oak tree*
 insect infestations in, 16–21, 23, 48
 mold growth in, 14–6, 21, 50, 62, 216, 229
 rain's impact on, 3, 9–10, 12–5, 48, 51–3
 reproductive cycle of, 154–8
Obama, Barack, 29
oceans, 216
 reflection of solar radiation using, 29, 98
 rising of, 5, 25–7, 33, 102, 113
offset programs, carbon, 184, 224, 248–9
oil. *See* fossil fuels
The Old Farmer's Almanac, 189–90
Olson, David, 195–7
ozone layer, 114–5

painted surfaces, solar reflection using, 113, 259
Pande, Rohini, 10, 70, 171
Pepco Energy, 172–4
Petrey, Scott, 219
photosynthesis, 56, 128, 130
phytophthora mold, 14–6, 21, 50, 62, 216, 229
pin oaks *(Quercus palustris),* 44, 47–9, 128, 154–8, 213–8, 242, 252
Politico, 166, 168
pollen, 47, 57–8, 101, 110
POWER Act, 202
Preston, Shaun, 62
Public Works Department, Takoma Park, 23, 50, 75

rain. *See* weather changes
Raskin, Jamie, 26, 69, 79–82, 89–94, 96, 159, 206–7, 245
Raskin, Tommy, 81, 90, 93, 207
red oaks *(Quercus falcata),* 1–4, 9–24, 51–2, 129, 132, 141, 241–2. *See also* oak trees; pin oaks
regenerative agriculture (carbon farming), 5, 46, 55–6, 71
ripraps, 85
rocks, carbon absorption by, 5, 55–6, 235, 259
Rogers, Fred, 118
RS Automotive, Takoma Park, 141–5, 244–5

Shut Down DC, 30–1
Sierra Club, 27, 136, 255
Silent Spring (Carson), 22
solar energy, 4, 20, 27, 32–5, 37, 71, 88, 125, 127, 134, 146, 148, 186, 190, 203, 206, 245, 259

INDEX

solar radiation modification
(SRM), 5, 55, 206, 235. *See also specific categories*
- atmosphere/stratosphere-based, 27, 72, 95–9, 103–21, 126, 149, 166–8, 236, 260
- cloud-based, 29, 46, 98, 103, 107, 113, 236, 259
- costs of, 46, 112–3, 260
- ethical considerations on, 114–21, 126, 167–8, 259
- land-based, 112–3, 259
- ocean-based, 29, 98
- risks of, 112, 114–5, 120, 166–8, 236–7, 259–60
- treaties/agreements on, 112–4, 166, 168

solutions to climate change. *See also specific solutions*
- carbon dioxide removal, 5, 27, 36, 38–9, 46–7, 52–68, 71–2, 76, 100, 108, 128, 148, 151–4, 181–6, 206, 209–13, 218–40, 241–2, 246–52, 259
- clean energy adoption, 4, 20, 27–39, 46–7, 57, 66, 71, 79–82, 88, 99, 108, 119, 124–7, 134, 141–145, 146–8, 168, 186, 190, 197–208, 211, 235, 245, 255, 259
- cost as factor in, 46, 55–6, 112–3, 225, 260
- solar radiation modification, 5, 27, 29, 46, 55, 72, 95–9, 103–21, 126, 149, 166–8, 206, 235–7, 259–60
- time as factor in, 27, 38–9, 46–7, 54–7, 206, 213, 232–40, 259–60

Stein, Dana, 37
sulfur dioxide, 27, 72, 96–9, 103–15, 126, 167, 236, 260

Tabor, Esther, 253, 261–2
Tabor, Mike, 253, 258, 261–2
Tajeddin, Yasmine, 152–3, 218–28
Takoma Park, Maryland/Willow Avenue, 25
- flooding in, 2, 13–4, 73–9, 80, 82–9, 92–4, 108, 199–200
- history of, 11–2
- tree loss on/around, 3–6, 9–10, 13–24, 38–9, 43–4, 47–53, 73, 91–3, 95, 128–9, 132–6, 138–41, 159, 161, 163, *169*, 169–74, 175–6, 186–7, 190–1, 200, 209–10, 216, 229, 232–3, 242–6, 252, 261–2

Takoma Park Presbyterian Church, 45, 73–9, 82, 116–21, 163, 187, 244
Talati, Shuchi, 168
Tallamy, Douglas, 129
Taylor, James, 37
temperature, 3–4, 21, 28–39, 44, 46, 52, 54, 57–8, 69–72, 95–101, 103–12, 114–6, 119–21, 128–9, 212–3, 216, 232–3, 242, 256. *See also* weather changes
- ice caps/sheets and, 2, 5, 14, 25–7, 47, 48, 102, 113, 153, 187, 257
- wildfires and, 147–8, 167

Thornhill, Elizabeth, 195
Thunberg, Greta, 142
Thwaites Glacier, 26
ticks, 2–3, 174–80
tree burial (geologic storage)
- in earth, 52–3, 58–68, 76, 100, 148, 151–4, 181–6, 209–13, 218–40, 241–2, 246–52
- in water, 249

trees. *See also* Takoma Park, Maryland/Willow Avenue
- burial, in land, of, 52–3, 58–68, 76, 100, 148, 151–4, 181–6, 209–13, 218–40, 241–2, 246–52
- burial, in water, of, 249
- carbon dioxide as fertilizer of, 127
- communication by, 12–3
- ecological contributions of, 3–4, 9–11, 65, 129, 242
- fungal infections in, 133

INDEX

gas leaks impacting, 129, 135–40
insect infestations in, 16–21, 23, 48
ivy's invasion of, 191–7
mold growth in, 14–6, 21, 50, 62, 216, 229
rain/storms impacting, 3, 9–10, 12–5, 48, 51–3, 91, 93, 128–9, 171, 216, 223–4, 229–31, 232
replanting of, 23, 156–7, 190, 258–9
reproductive cycles of, 48–9, 213–5, 217–8
root systems of, 12–6, 92–3
temperature's impact on, 47–9, 69–71, 128–9
wildlife's symbiosis with, 3, 9–10, 129, 242
Trump, Donald, 80–1, 90, 92

United Nations, 4, 51, 56–7, 66, 234, 237
University of Maryland, 23, 35, 49, 105, 150–3
US Forest Service, 66

VLA15 vaccine, 180
volcanoes, 95–9, 103, 106–7, 109, 131, 236

Wang, Jake, 152–3
Washington Gas Company, 135, 138–9
Washington Post, 28, 57, 83, 171
weather changes. *See also* climate change; flooding
  and droughts, 9, 12–3, 18, 20, 27, 38, 47–8, 51, 88, 101, 146, 153, 156–7, 164, 210, 215–7, 256
  and excessive rainfall/storms, 3, 9–10, 12–5, 18, 20–1, 27, 38–9, 48, 51–3, 62, 66, 72–9, 80, 82–9, 92–4, 107, 115, 169–72, 177, 187, 210, 216, 223–4, 229–31, 232, 256, 260
  and icecaps/ocean levels, 2, 5, 14, 25–7, 47, 48, 102, 113, 153, 187, 257
  temperature-based, 2–5, 14, 21, 25–39, 44, 46–8, 52, 54, 57–8, 69–72, 95–116, 119–21, 128–9, 147–8, 153, 167, 187, 212–3, 216, 232–3, 242, 256–7
White House Office of Science and Technology Policy, 149, 166
white oaks, 1–2, 10, 12–23, 50–1, 62, 181–3, 232, 258. *See also* oak trees
wildfires, 5, 48, 51, 65–6, 109, 242, 257
  Canadian, 131–2, 135, 146–53, 157, 160, 177, 233, 236, 256
  mulch-based, 210–1
Willow Avenue. *See* Takoma Park, Maryland/Willow Avenue
willow oaks, 132–3, 140–1, 159, 171, 175, 187, 190–1
wind energy, 27, 34–7, 80, 125, 127, 146, 148, 190, 245
  ocean-based, 197–208

XPRIZE, 60, 71

Yuan, Tianle, 98, 103, 105–13, 115, 206, 235

Zachariae Isstrøm ice sheet, 102
Zahradka, George, 227
Zahradka, Libby, 225–9, 231, 233
Zeng, Annarita, 226, 232, 246, 250
Zeng, Elisa, 235
Zeng, Luca, 238
Zeng, Ning, 43, 49–53, 57–68, 71–2, 76, 95–102, 105–6, 148–54, 181–6, 206–7, 209–14, 218–40, 241, 246–52

ABOUT THE AUTHOR

MIKE TIDEWELL is a writer and climate activist living in Takoma Park, Maryland. His books include *Bayou Farewell*, about the disappearing wetlands and Cajun culture in south Louisiana. As a contributing travel writer for *The Washington Post,* Tidwell has won four Lowell Thomas Awards, the highest prize in American travel journalism. In 2002 he founded the Chesapeake Climate Action Network, and still serves as executive director, where he has led local, state, and national campaigns for clean energy. He lives on Willow Avenue in Takoma Park with his wife, Beth, and their cat, Macy Gray.